# Triumph Pre-unit Construction Twins Owners Workshop Manual

by Jeff Clew
**Member of the Guild of Motoring Writers**

**Models covered** Covers all 500 cc and 650 cc models fitted with a separate engine and gearbox from 1947 — 1962

5T   Speed Twin
6T   Thunderbird
Tiger 100
T110
T120 Bonneville
TR5 Trophy (500 cc)
TR6 Trophy (650 cc)

ISBN **978 0 85696 251 6**

ABC

2

Printed in Malaysia   *(251–4P6)*

**Haynes Publishing**
Sparkford, Yeovil,
Somerset BA22 7JJ, England

**Haynes North America, Inc**
859 Lawrence Drive,
Newbury Park,
California 91320, USA

# Acknowledgements

Our thanks are due to Norton-Triumph International Limited for their assistance. Brian Horsfall gave the necessary assistance with the overhaul and devised ingenious methods for overcoming the lack of service tools. Les Brazier took the photographs that accompany the text.

We thank the Avon Rubber Company Limited for their advice on tyre fitting and NGK Spark Plugs (UK) Limited for the provision of spark plug photographs.

# About this manual

The author of this manual has the conviction that the only way in which a meaningful and easy-to-follow text can be written is to carry out the work himself, under conditions similar to those found in the average household. As a result, the hands seen in the photographs are those of the author. Even the machines are not new: examples which have covered a considerable mileage are selected, so that the conditions encountered would be typical of those encountered by the average rider/owner. Unless specially mentioned, and therefore considered essential, Triumph service tools have not been used. There are invariably alternative means of slackening or removing some vital component when service tools are not available, but risk of damage is to be avoided at all costs.

Each of the eight chapters is divided into numbered sections. Within the sections are numbered paragraphs. Cross-reference throughout the manual is quite straightforward and logical. For example, when reference is made 'See Section 6.2' it means section 6, paragraph 2 in the same chapter. If another chapter were meant, the reference would read (See Chapter 2, section 6.2'. All photographs are captioned with a section/paragraph number to which they refer, and are always relevant to the chapter text adjacent.

Figure numbers (usually line illustrations) appear in numerical order, within a given chapter. Fig 1.1 therefore refers to the first figure in Chapter 1. Left hand and right hand descriptions of the machines and their component parts refer to the left and right when the rider is seated, facing forward.

Motorcycle manufacturers continually make changes to specifications and recommendations, and these, when notified, are incorporated into our manuals at the earliest opportunity.

Whilst every care is taken to ensure that the information in this manual is correct no liability can be accepted by the authors or publishers for loss, damage or injury caused by any errors in or omissions from the information given.

# Introduction to the Triumph 500 cc and 650 cc Pre-unit construction vertical twins

Contrary to popular belief, the first Triumph vertical twin engine was designed and manufactured as far back as 1914, although a complete machine was not built. It was not until 1933 that another twin cylinder design emerged, this time in the form of a unit-construction 650 cc engine with geared primary drive, designed by Val Page. One of the completed models, with sidecar attached, covered 500 miles in 500 minutes at Brooklands immediately after the outfit had competed successfully in that year's International Six Days Trial. This feat, which was observed by the Auto-Cycle Union throughout, won the Maudes Trophy for Triumph Motors.

In January 1936 the Triumph Engineering Company Limited was formed to take over the motor cycle manufacturing activities of Triumph Motors. A new designer, Edward Turner, was appointed and it was Turner who inspired the Speed Twin model that had sensational impact on the motor cycle world during 1937. Indeed, this is the model that is the true ancestor of today's vertical twin designs and the one that established an entirely new trend in motor cycling.

The first of the modern 650 cc models was the Thunderbird, first manufactured during 1949. Although by no means a sluggard, there was evidence of demand for a model with even higher performance and in 1953 the T110 model was announced, the coding relating to the anticipated performance in miles per hour. The T110 was one of the first models to feature swinging arm rear suspension; the earlier Thunderbird models were supplied with either a rigid frame or a sprung hub of Triumph manufacture - another Triumph innovation.

In due course the T110 was supplemented by another model capable of even higher performance, the T120 of 1958. This latter machine was named the Bonneville in recognition of Johnny Allen's record breaking attempts in America with an unsupercharged 650 cc engine encased in a cigar-like shell. The attempts took place on the Bonneville Salt Flats on 25th September, 1955 when Allen achieved a mean speed of 193.72 mph over the flying kilometer and 192.30 mph over the measured mile. These were the highest speeds ever covered on a motor cycle at that time and it is unfortunate that this remarkable achievement was somewhat clouded by squabbles at International level about the official recognition of this feat.

Production of the Bonneville model continued virtually unchanged until the end of the 1962, when it was announced that the 1963 models would be built on the unit-construction principle, to bring them in line with the other models then in the current range.

Mention should also be made of the Trophy model which started life in 1948 as a competition model, for trials and scrambles use. Early models were of 500 cc capacity only they used a development of the square-finned generator engine that was manufactured during the war. In 1951 this engine was superseded by a new close-pitch finned engine of all alloy construction, used also in modified form of the Tiger 100 model. A 650 cc Trophy model was added to the range in 1956 and eventually the 500 cc model was phased out.

The Triumph engine has an outstanding reputation for reliability and high performance, two attributes that do not often go hand in hand. This is why the Triumph twin engine is invariably the one chosen by builders of 'specials', who wish to create their own machine by marrying together a collection of parts not necessarily from the one factory.

# Contents

**1952 Triumph Tiger 100 - left-hand side**

**1952 Triumph Tiger 100 - right-hand side**

# Safety first!

Professional motor mechanics are trained in safe working procedures. However enthusiastic you may be about getting on with the job in hand, do take the time to ensure that your safety is not put at risk. A moment's lack of attention can result in an accident, as can failure to observe certain elementary precautions.

There will always be new ways of having accidents, and the following points do not pretend to be a comprehensive list of all dangers; they are intended rather to make you aware of the risks and to encourage a safety-conscious approach to all work you carry out on your vehicle.

### Essential DOs and DON'Ts

**DON'T** start the engine without first ascertaining that the transmission is in neutral.

**DON'T** suddenly remove the filler cap from a hot cooling system – cover it with a cloth and release the pressure gradually first, or you may get scalded by escaping coolant.

**DON'T** attempt to drain oil until you are sure it has cooled sufficiently to avoid scalding you.

**DON'T** grasp any part of the engine, exhaust or silencer without first ascertaining that it is sufficiently cool to avoid burning you.

**DON'T** allow brake fluid or antifreeze to contact the machine's paintwork or plastic components.

**DON'T** syphon toxic liquids such as fuel, brake fluid or antifreeze by mouth, or allow them to remain on your skin.

**DON'T** inhale dust – it may be injurious to health (see *Asbestos* heading).

**DON'T** allow any spilt oil or grease to remain on the floor – wipe it up straight away, before someone slips on it.

**DON'T** use ill-fitting spanners or other tools which may slip and cause injury.

**DON'T** attempt to lift a heavy component which may be beyond your capability – get assistance.

**DON'T** rush to finish a job, or take unverified short cuts.

**DON'T** allow children or animals in or around an unattended vehicle.

**DON'T** inflate a tyre to a pressure above the recommended maximum. Apart from overstressing the carcase and wheel rim, in extreme cases the tyre may blow off forcibly.

**DO** ensure that the machine is supported securely at all times. This is especially important when the machine is blocked up to aid wheel or fork removal.

**DO** take care when attempting to slacken a stubborn nut or bolt. It is generally better to pull on a spanner, rather than push, so that if slippage occurs you fall away from the machine rather than on to it.

**DO** wear eye protection when using power tools such as drill, sander, bench grinder etc.

**DO** use a barrier cream on your hands prior to undertaking dirty jobs – it will protect your skin from infection as well as making the dirt easier to remove afterwards; but make sure your hands aren't left slippery. Note that long-term contact with used engine oil can be a health hazard.

**DO** keep loose clothing (cuffs, tie etc) and long hair well out of the way of moving mechanical parts.

**DO** remove rings, wristwatch etc, before working on the vehicle – especially the electrical system.

**DO** keep your work area tidy – it is only too easy to fall over articles left lying around.

**DO** exercise caution when compressing springs for removal or installation. Ensure that the tension is applied and released in a controlled manner, using suitable tools which preclude the possibility of the spring escaping violently.

**DO** ensure that any lifting tackle used has a safe working load rating adequate for the job.

**DO** get someone to check periodically that all is well, when working alone on the vehicle.

**DO** carry out work in a logical sequence and check that everything is correctly assembled and tightened afterwards.

**DO** remember that your vehicle's safety affects that of yourself and others. If in doubt on any point, get specialist advice.

**IF,** in spite of following these precautions, you are unfortunate enough to injure yourself, seek medical attention as soon as possible.

### Asbestos

Certain friction, insulating, sealing, and other products – such as brake linings, clutch linings, gaskets, etc – contain asbestos. *Extreme care must be taken to avoid inhalation of dust from such products since it is hazardous to health*. If in doubt, assume that they *do* contain asbestos.

### Fire

Remember at all times that petrol (gasoline) is highly flammable. Never smoke, or have any kind of naked flame around, when working on the vehicle. But the risk does not end there – a spark caused by an electrical short-circuit, by two metal surfaces contacting each other, by careless use of tools, or even by static electricity built up in your body under certain conditions, can ignite petrol vapour, which in a confined space is highly explosive.

Always disconnect the battery earth (ground) terminal before working on any part of the fuel or electrical system, and never risk spilling fuel on to a hot engine or exhaust.

It is recommended that a fire extinguisher of a type suitable for fuel and electrical fires is kept handy in the garage or workplace at all times. Never try to extinguish a fuel or electrical fire with water.

**Note:** *Any reference to a 'torch' appearing in this manual should always be taken to mean a hand-held battery-operated electric lamp or flashlight. It does **not** mean a welding/gas torch or blowlamp.*

### Fumes

Certain fumes are highly toxic and can quickly cause unconsciousness and even death if inhaled to any extent. Petrol (gasoline) vapour comes into this category, as do the vapours from certain solvents such as trichloroethylene. Any draining or pouring of such volatile fluids should be done in a well ventilated area.

When using cleaning fluids and solvents, read the instructions carefully. Never use materials from unmarked containers – they may give off poisonous vapours.

Never run the engine of a motor vehicle in an enclosed space such as a garage. Exhaust fumes contain carbon monoxide which is extremely poisonous; if you need to run the engine, always do so in the open air or at least have the rear of the vehicle outside the workplace.

### The battery

Never cause a spark, or allow a naked light, near the vehicle's battery. It will normally be giving off a certain amount of hydrogen gas, which is highly explosive.

Always disconnect the battery earth (ground) terminal before working on the fuel or electrical systems.

If possible, loosen the filler plugs or cover when charging the battery from an external source. Do not charge at an excessive rate or the battery may burst.

Take care when topping up and when carrying the battery. The acid electrolyte, even when diluted, is very corrosive and should not be allowed to contact the eyes or skin.

If you ever need to prepare electrolyte yourself, always add the acid slowly to the water, and never the other way round. Protect against splashes by wearing rubber gloves and goggles.

### Mains electricity and electrical equipment

When using an electric power tool, inspection light etc, always ensure that the appliance is correctly connected to its plug and that, where necessary, it is properly earthed (grounded). Do not use such appliances in damp conditions and, again, beware of creating a spark or applying excessive heat in the vicinity of fuel or fuel vapour. Also ensure that the appliances meet the relevant national safety standards.

### Ignition HT voltage

A severe electric shock can result from touching certain parts of the ignition system, such as the HT leads, when the engine is running or being cranked, particularly if components are damp or the insulation is defective. Where an electronic ignition system is fitted, the HT voltage is much higher and could prove fatal.

# Ordering spare parts

When ordering spare parts for any of the Triumph pre-unit-construction vertical twins, it is advisable to deal direct with an official Triumph agent who may still be able to supply many of the items ex-stock. Parts cannot be obtained direct from the Norton-Triumph International Limited; all orders must be routed through an approved agent, even if the parts required are not held in stock.

Always quote the engine and frame numbers in full. Include any letters before or after the number itself. The frame number will be found stamped on the left hand front down tube, adjacent to the steering head. The engine number is stamped on the left hand crankcase, immediately below the base of the cylinder barrel.

Use only parts of genuine Triumph manufacture. Pattern parts are available but in many instances they will have an adverse effect on performance and/or reliability. Some complete units may be available on a 'service exchange' basis from the manufacturer or even the dealer's own service affording an economic method of repair without having to wait for parts to be reconditioned. Details of the parts available, which include petrol tanks, front forks, front and rear frames, clutch plates, brake shoes etc. can be obtained from any Triumph agent. It follows that the parts to be exchanged must be acceptable before factory reconditioned replacements can be supplied.

Retain any broken or worn parts until a new replacement has been obtained. Often these parts are required as a pattern for identification purposes, a problem that becomes more acute when a machine is classified as obsolete. In an extreme case, where replacements are not available, it may be possible to reclaim the original or to use it as a pattern for having a replacement made. Many older machines are kept on the road in this way, long after a manufactuer's spares have ceased to be available.

Some of the more expendable parts such as spark plugs, bulbs, tyres, oils and greases etc., can be obtained from accessory shops and motor factors, who have convenient opening hours. charge lower prices and can often be found not far from home. It is also possible to obtain parts on a Mail Order basis from a number of specialists who advertise regularly in the motor cycle magazines.

1  Engine number location

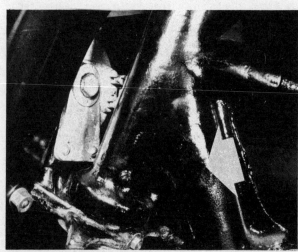

2  Frame number location

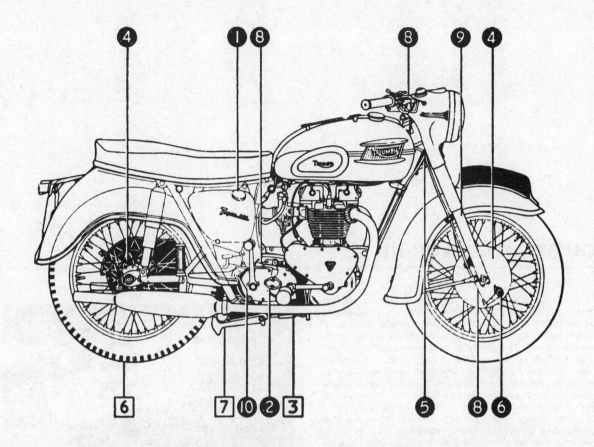

**Lubrication chart key**

| | |
|---|---|
| 1 Oil tank | 6 Brake cam spindles |
| 2 Gearbox | 7 Brake pedal spindle |
| 3 Parimary chaincase | 8 Control cables |
| 4 Wheel bearings | 9 Telescopic forks (damping fluid fork oil) |
| 5 Steering head bearings | 10 Swinging arm fork spindle |

Circles: lubrication points on right-hand side
Squares: lubrication points on left-hand side

# Recommended lubricants

| Component | Type of lubricant | Castrol Grade |
|---|---|---|
| Engine ... ... ... ... ... ... | Summer : 20W/50 Multigrade ... ... ... ... ... | Castrol GTX |
| | Winter: 10W/30 Multigrade ... ... ... ... ... | Castrolite |
| Gearbox ... ... ... ... ... | SAE 50 Monograde ... ... ... ... ... | Castrol Grand Prix |
| Primary chaincase ... ... ... ... | SAE 20 ... ... ... ... ... | Castrolite |
| All grease points: Brake cam spindles etc. ... ... ... | Lithium based high melting point ... ... ... ... ... | Castrol LM Grease |
| All lubrication points; ... ... ... including front forks and cables | Light oil, 10W/30 Multigrade (forks SAE 20) ... ... ... | Castrolite |

# Routine maintenance

Periodic routine maintenance is a continuous process that commences immediately the machine is used. It must be carried out at specified mileage recordings or on a calendar date basis if the machine is not used regularly, whichever falls soonest. Maintenance should be regarded as an insurance policy, to help keep the machine in peak condition and to ensure long, trouble-free service. It has the additional benefit of giving early warning of any faults that may develop and will act as a regular safety check, to the obvious advantage of both rider and machine alike.

The various maintenance tasks are described under their respective mileage and calendar headings. Accompanying diagrams are provided, where necessary. It should be remembered that the interval between the various maintenance tasks serves only as a guide. As the machine gets older or is used under particularly adverse conditions, it would be advisable to reduce the period between each check.

Some of the tasks are described under their respective mileage and calendar headings. Accompanying diagrams are provided, where necessary. It should be remembered that the interval between the various maintenance tasks serves only as a guide. As the machine gets older or is used under particularly adverse conditions, it would be advisable to reduce the period between each check.

Some of the tasks are described in detail, where they are not mentioned fully as a routine maintenance item in the text. If a specific item is mentioned, but not described in detail, it will be covered fully in the appropriate chapter. No special tools are required for the normal routine maintenance tasks. The tools contained in the tool kit supplied with every new machine will prove adequate for each task or if they are not available, the tools found in the average household will usually suffice.

RM1 Check oil tank content frequently and top up when necessary

## Weekly or every 250 miles (400 km)

Check level in oil tank and top up if necessary.
Check level in primary chaincase and lubricate the rear chain.
Check battery acid level and tyre pressures.
Check chaincase cover screws

## Monthly or every 1000 miles (1600 km)

Check brake action and adjust.
Change oil in primary chaincase.
Lubricate the control cables (see accompanying diagram) and grease the swinging arm fork pivot (grease nipple on underside).
Remove, clean and re-lubricate the final drive chain.

RM2 Do not neglect the gearbox. Correct oil content is essential

# Routine Maintenance

Check all nuts, bolts etc for tightness.
Adjust tension of primary and final drive chains.
Check gearbox oil and replenish
Check gearbox clamp bolts.
Add few drops thin oil to SU carburettor dashpot (6T model only) and to the distributor.

## Six weekly or every 1500 miles (2400 km)

Dismantle and clean carburettor
Clean and re-oil air filter element
Drain oil tank when warm and refill
Clean oil filters.

## Three monthly or every 3000 miles (4800 km)

Check and adjust valve clearances
Clean and adjust spark plugs.
Check wheel bearings.
Check and adjust contact breaker points.

## Six monthly or every 6000 miles (9600 km)

Drain gearbox oil and refill.
Examine front forks for oil leakage. Drain and refill.
Grease brake pedal spindle.
Check ignition timing.
Check play in head races and adjust.
Lubricate contact breaker.

## Yearly or every 12,000 miles (19,200 km)

Decarbonise and top overhaul.
Grease wheel and steering head bearings.

These latter two tasks will necessitate a certain amount of dismantling, details of which are given in Chapters 5 and 6.

It should be noted that even when six monthly and yearly maintenance tasks have to be undertaken, the weekly, monthly, six weekly and three monthly services must be completed. There is no stage at any point during the life of the machine when a routine maintenance task can be ignored.

### Routine maintenance and capacities data

| | |
|---|---|
| Engine | Oil tank capacity<br>5 pints (2.84 litres) |
| Gearbox | 2/3 pint (380 ccs) |
| Front forks<br>(per leg) | 1/6 pint (95 ccs) - all 1947 - 59 models<br>1/4 pint (142 ccs) - 1960 - 62 solo models<br>3/8 pint (213 ccs) - 1960 - 62 models with longer lower fork legs (sidecar) |
| Primary chaincase | 1/3 pint (190 ccs) |
| Contact breaker gap | 0.012 in. (magneto)<br>0.014 in. (coil) |
| Spark plug gap | 0.020 in. (magneto)<br>0.025 in. (coil) |

RM3 Make sure chaincase oil is of recommended viscosity

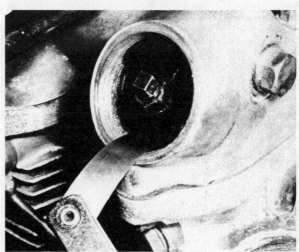

RM4 Check valve clearances by removing rocker box caps

| | |
|---|---|
| Tyre pressures | See Specifications, Chapter 7. |

Mention has not been made of the lighting equipment, horn and speedometer, all of which must be checked frequently to ensure they are in good working order. The tyres also should be checked regularly and renewed if they wear unevenly, have splits or cracks in the side walls or if the depth of tread approaches the statutory minimum. Neglect of any of these points may render the owner liable to prosecution, apart from creating a safety hazard.

# Chapter 1 Engine

**Contents**

**Specifications**

| Model No. | 5T | T100 | TR5 | 6T | TR6 | T110 | T120 |
|---|---|---|---|---|---|---|---|
| Name | Speed Twin | Tiger 100 | Trophy 500 | Thunderbird | Trophy 650 | — | Bonneville |
| Capacity ccs | 498 | 498 | 498 | 649 | 649 | 649 | 649 |
| Bore mm | 63 | 63 | 63 | 71 | 71 | 71 | 71 |
| Stroke mm | 80 | 80 | 80 | 82 | 82 | 82 | 82 |
| Compression ratio | 7 : 1 | 8 : 1* | 8 : 1* | 7 : 1 | 8 : 1* | 8 : 1 | 8.5 : 1 |
| BHP @ rpm | 27 @ 6,300 | 34 @ 6300 | 33 @ 6500 | 34 @ 6300 | 40 @ 6500 | 40 @ 6500 | 46 @ 6500 |
| Engine sprocket, No. of teeth | | | | | | | |
|   Solo | 22 | 22 | 21 | 24 | 24 | 24 | 24 |
|   Sidecar | 19 | 19 | — | 21 | — | 21 | 21 |

Valve clearances ...        ...        ...        ...        ...        ...        ...        ...                        See Section 35

Valve timing
with 0.020 in
valve clearances

| | | | | | |
|---|---|---|---|---|---|
| Inlet opens ° BTDC | 26.5 | 27 | as 5T | as TR5 | as TR5 | 34 |
| Inlet closes ° ABDC | 69.5 | 48 | | | | 55 |
| Exhaust opens ° BBDC | 61.5 | 48 | | | | 48 |
| Exhaust closes ° ATDC | 35.5 | 27 | | | | 27 |

+  iron engine 0.001 in.                                              +lower on earlier models

*Note:*      *The above data applies mainly to 1956 - 9 models. It is advisable to cross-check the data relating to tappet clearances and valve timing with the manufacturer's original publication since the use of different camshafts may call for variations in the settings given.*

## 1  General description

The engine fitted to the Triumph pre-unit construction vertical twins makes extensive use of aluminium alloy castings, especially the engines that have more sporting characteristics. On the very early models, cast iron cylinder blocks and cylinder heads formed part of the basic specification, but from 1951 onwards models such as the Tiger 100 and Trophy Twin employed a new design of die cast aluminium alloy cylinder head and cylinder block, characterised by the close pitch finning. Some models, such as the Speed Twin and the Thunderbird retained the cast iron components until a much later date. Others, such as the T110, used a combination of a cast iron cylinder block and a light alloy cylinder head.

Whenever a light alloy cylinder head is used, cast-in austenitic valve seats are employed. The overhead valves are actuated by rocker arms enclosed within detachable alloy rocker boxes. The push rods are also of aluminium alloy, fitted with hardened end pieces. The tappet guide blocks are a press fit in the cylinder block.

'H' section connecting rods of hinduminium alloy, with detachable caps and steel-backed shell bearings, carry aluminium alloy die-cast pistons, each with two compression rings and one oil scraper ring. The two-throw crankshaft has a detachable shrunk-on cast iron flywheel, retained in a central position by three high tensile steel bolts. Earlier engines have a two piece built-up crankshaft assembly, and smaller diameter main bearings. Engines fitted with the later, larger diameter main bearings can be identified by the bulge of the bearing housing on the timing side of the crankcase, to the rear of the timing chest. The change in bearing specification took place around 1955 and if any attempt is to be made to tune the engine, it should be applied only to one with the larger bearings.

The separate inlet and exhaust camshafts operate in sintered bronze bushes, mounted transversely in the upper part of the crankcase. The camshafts are driven by the train of timing gears, from the right hand end of the crankshaft. The inlet camshaft provides the drive for the oil pump and the rotary breather valve disc, whilst the exhaust camshaft drives the contact breaker and, on some models, the tachometer drive gearbox.

Power from the engine is transmitted in the conventional manner through the engine sprocket and primary chain to the clutch unit. A shock absorber of the spring and cam type is located on the drive side (left hand) end of the crankshaft. On models fitted with an alternator, this arrangement is not practicable and the shock absorber is built into the clutch centre.

## 2  Operation with engine in frame

It is not necessary to remove the engine unit from the frame unless the crankshaft assembly and/or main bearings require attention. Most operations can be accomplished with the engine in place, such as:
1  Removal and replacement of cylinder head (rocker boxes may be removed separately where necessary clearance exists)
2  Removal and replacement of cylinder barrel and pistons.
3  Removal and replacement of the magneto or alternator.
4  Removal and replacement of primary drive components.
5  Removal and replacement of oil pump.

When several operations need to be undertaken simultaneously, for example during a major overhaul, it will probably be advantageous to remove the complete engine unit from the frame, an operation that should take approximately 2 to 2½ hours. This is a two man job; the engine unit weighs 135 lbs.

## 3  Operations with engine removed

1  Removal and replacement of the main bearings.
2  Removal and replacement of the crankshaft assembly.
3  Removal and replacement of the camshafts.
4  Renewing the camshaft and timing gear bushes.

## 4  Method of engine removal

Because the machine has a separate engine and gearbox, it is possible to remove either the engine or the gearbox alone, without having to disturb the other assembly. A certain amount of dismantling is necessary in order to gain full access to either component, but this is of a superficial nature.

## 5  Removing the engine unit

1  Place the machine on the centre stand and make sure that it is standing firmly, on level ground. On older machines of the rigid frame type, the rear stand will have to be used for this purpose.
2  Turn off the petrol taps and disconnect the petrol pipes by unscrewing the union nut joints.
3  Remove the petrol tank. The fixing arrangement will vary according to the year of manufacture of the machine. Early models have four bolts that thread into inserts in the base of the tank, which may be wired together for security. Later models have a retaining strap secured by a nut and bolt at the front and a rear cross bolt. Do not lose any of the rubber buffers or insulating washers that may be displaced as the tank is removed. Note that on the nacelle headlamp models, it will be necessary to remove the screws retaining the rear of the nacelle, to give sufficient clearance.

5.5 Finned clamps retain exhaust pipes to cylinder head stubs

5.5a Silencers bolt to tapped lug on frame

5.5b If desired, silencers can be detached by slackening clamp bolt

5.6 Head steady bolts clamp on down tube

5.8 Exercise care or thin pipes will 'neck' if twisted

5.9 Detach igniton advance cable at handlebar control end

5.11 Dynamo will pull out as a complete unit

5.12 Left-hand footrest seats on outer chaincase cover

5.14 Whole primary transmission assembly must be removed

4  Detach the battery leads, so that the battery is isolated. If necessary, the battery can be removed at this stage and bench charged so that it is available in a fully charged condition during the rebuild.

5  Slacken the finned clamps around the exhaust pipes and detach the pipes from the stays that bolt to the front engine mounting or crankcase stud. Drive the exhaust pipes off the exhaust port stubs with a rawhide mallet. Remove the complete exhaust system.

6  Detach the cylinder head torque stays by removing the two nuts and the bolt that passes through the frame lug. A slightly different arrangement is employed on the T120 Bonneville model, the stay taking the form of a flat plate that supports the carburettor float chamber. In this case the plate, float chamber and petrol pipes should be removed as a complete unit.

7  Remove the air cleaner hose and unscrew the flange nuts that retain the carburettor(s) to the cylinder head. Tie the carburettor(s) to a frame tube so that it is out of the way and unlikely to be damaged as the engine is lifted out.

8  Unscrew the acorn nuts that retain the thin rocker feed pipe to the rocker box unions. Care is necessary when slackening the nuts because if the unions themselves turn at the same time, the feed pipe will 'neck' and restrict the oil flow when the engine is re-used. Use a spanner on the flats of the unions to restrain them from movement, if necessary.

9  If a magneto is fitted, detach the advance/retard control cable from the handlebar control and coil it neatly in close proximity to the magneto. Detach the cut-out button lead from the end cover of the magneto, if fitted. If a distributor is fitted, it will be necessary to detach the electrical leads.

10  To remove the dynamo as a complete unit, slacken the two screws that pass through the retaining strap on the front of the right-hand crankcase. Remove the acorn nut from the forward end of the timing cover. Detach the electrical leads from the left-hand end of the dynamo; they are retained by a small crescent-shaped insulating strip, held in place by a central screw. The screw will remain captive with the insulating strip.

11  Pull the dynamo out of position from the left-hand side, rotating it if necessary to clear the back of the timing chest. There should be no need to remove the dynamo drive pinion as there is sufficient clearance for it to pass through. Note the use of a cork gasket, to effect a seal between the dynamo body and the back of the timing chest.

12  Remove both footrests, leaving the footrest rod in position. Take out the split pin retaining the operating rod to the brake pedal. Unscrew the pedal spindle nut and withdraw the pedal. Do not lose the small clevis pin through the trunnion on the end of the brake operating rod.

13  Place a drip tray under the primary chaincase and slacken and remove the retaining screws around the periphery of the chaincase. This will release the oil content. Ease the outer cover away gently, especially when an alternator is fitted, otherwise there is risk of damaging the stator windings.

14  Before the back half of the chaincase can be removed, it will be necessary to dismantle the primary transmission. On magneto ignition models, commence by slackening and unscrewing the nut on the end of the engine shaft shock absorber. Bend back the tab washer and jar the nut loose with a socket spanner and mallet. If necessary, the engine can be locked by engaging top gear and applying the rear brake. When the nut has been unscrewed, take off the collar in which it seats, the shock absorber spring and the front portion of the shock absorber cam. If the primary chain is fitted with a spring link, this can be removed, together with the chain itself and the engine sprocket.

15  If the engine is of the alternator type, it is first necessary to remove the stator coil assembly. Detach the snap connectors to the stator coil which will be found in the vicinity of the rear of the chaincase, then remove the three nuts and washers that retain the coil assembly to the studs around the rotor. Draw the disconnected stator leads through the rear of the chaincase, if necessary removing the sleeve nut that threads into the chaincase tunnel, to give sufficient clearance. The whole assembly can then

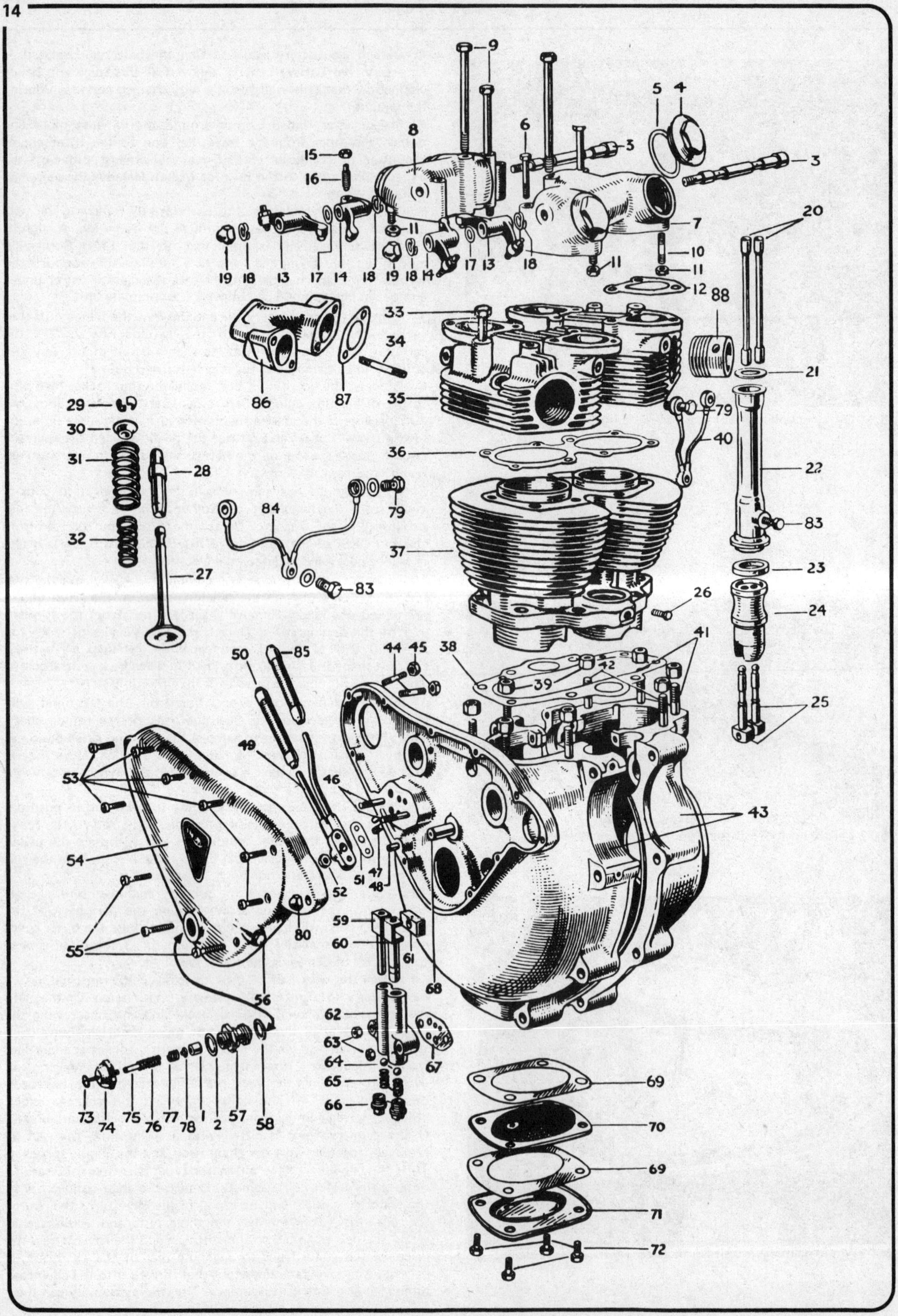
9
9
5
4
8
6
3
3
15
16
11
7
20
10
11
11
19
18
13
17
14
18
19
18
14
17
13
18
12
88
33
34
21
86
87
35
79
29
30
36
40
31
28
22
84
79
83
32
37
27
83
23
26
24
50
85
44
45
38
41
42
39
25
49
53
46
43
54
51
47
48
52
80
59
60
61
55
68
56
62
63
67
64
65
66
73
74
75
76
77
78
1
2
57
58
69
70
69
71
72

**Fig. 1.1 Crankcase, cylinder block and cylinder head assembly - early T100 and TR5 models**

1 Pressure release valve piston
2 Pressure release valve washer
3 Rocker spindle - 2 off
4 Rocker box inspection cap - 4 off
5 Inspection cap washer - 4 off
6 Rocker box bolt - 4 off
7 Exhaust rocker box
8 Inlet rocker box
9 Cylinder head bolt (long - 2 off
10 Rocker box stud - 4 off
11 Nut - 4 off
12 Rocker box gasket - 4 off
13 Rocker arm (left-hand) - 2 off
14 Rocker arm (right-hand) - 2 off
15 Tappet locknut - 4 off
16 Rocker adjusting pin - 4 off
17 Rocker arm thrust washer (½ inch) - 6 off
18 Rocker arm thrust washer (3/8 inch) - 2 off
19 Domed nut - 2 off
20 Push rods - 4 off
21 Push rod tube washer - 2 off
22 Push rod tube - 2 off
23 Push rod tube washer (thick) - 2 off
24 Tappet guide block - 2 off
25 Tappet - 4 off
26 Tappet guide block locking screw - 2 off
27 Valve - 4 off
28 Valve guide - 4 off
29 Collets - 8 off
30 Valve spring cap - 4 off
31 Outer valve spring - 4 off
32 Inner valve spring - 4 off
33 Cylinder head bolt (short) - 4 off
34 Carburettor stud - 2 off
35 Cylinder head
36 Cylinder head gasket
37 Cylinder block
38 Cylinder base gasket
39 Cylinder base nut - 8 off
40 Inlet oil drain pipe
41 Crankcase stud - 6 off
42 Crankcase stud dowl - 2 off
43 Crankcase
44 Magneto fixing stud - 3 off
45 Nut - 3 off
46 Oil pump fixing stud - 2 off
47 Oil junction block stud
48 Dowel
49 Oil junction block
50 Flexible oil pipe - 2 off
51 Junction block gasket
52 Nut
53 Timing cover screw (medium) - 7 off
54 Timing cover
55 Timing cover screw (long) - 3 off
56 Plug
57 Pressure release valve body
58 Washer
59 Oil pump feed plunger
60 Oil pump scavenge plunger
61 Oil pump slider block
62 Oil pump body
63 Nut - 2 off
64 Oil pump ball valve - 2 off
65 Oil pump ball valve spring - 2 off
66 Oil pump valve plug - 2 off
67 Oil pump gasket
68 Intermediate timing gear spindle
69 Crankcase oil filter gasket
70 Oil filter gauze
71 Crankcase oil filter cover
72 Crankcase oil filter bolt - 4 off

73 Indicator button
74 Pressure release valve cap
75 Indicator rubber tube
76 Spring (main)
77 Spring (auxilliary)
78 Indicator shaft
79 Drain pipe adaptor (Banjo union bolt) - 4 off
80 Dome nut
83 Adaptor for push rod tube (Banjo union bolt) - 2 off
84 Exhaust oil drain pipe
85 Oil pipe connection
86 Inlet manifold
87 Inlet manifold gasket
88 Exhaust pipe adaptor - 2 off

*Note: The quantity of some parts used may vary, according to the model designation*

be pulled off the studs and placed well clear of the machine, taking care to prevent damage to the coil windings. To remove the rotor, bend back the tab washer and unscrew the retaining nut. It can be jarred loose with a socket spanner and mallet. The rotor is keyed on to the crankshaft and should pull off quite easily.

16 It will be necessary to take off the engine sprocket and the clutch assembly in unison, since engines of the alternator type employ an endless primary chain. This will also apply to engines with magneto ignition that have a similar type of chain. Commence operations by slackening off the four adjuster nuts in the clutch pressure plate, using a penknife or some similar device to depress the clutch springs whilst the nuts are unscrewed. This is to free the springs from the location pip on the underside of each nut which acts as a self-locking device. A slotted screwdriver will need to be used for the initial slackening.

17 Remove the adjusting nuts completely, then withdraw the pressure plate complete with the clutch springs and the cups in which they seat. The individual clutch plates, both plain and inserted can now be drawn out, using two pieces of wire with a hook at the end.

18 When all the plates are withdrawn, access is available to the inner drum and shock absorber unit. With the engine still locked in position and a locking bar between the inner and outer clutch drums, unscrew the centre retaining nut and remove it, together with the cupped washer.

19 The complete clutch assembly can now be pulled off the clutch centre by inserting the appropriate Triumph service tool and screwing it home fully so that the full depth of thread engages. When the centre bolt is tightened, the clutch will be drawn off the mainshaft taper.

20 If the service tool is not available, the clutch pressure plate can be utilised to good effect. Refit it, without the clutch plates or the cups and springs, using the adjuster nuts with washers placed beneath them. (It will be necessary to file a notch in each washer, to accommodate the pip on the underside of each adjuster nut.) Unscrew the adjuster locknut in the centre of the clutch pressure plate and screw the adjuster inwards. It should pull the clutch assembly off the clutch centre splines when the end of the adjuster abuts on the end of the mainshaft.

21 The above techniques should NOT be used if the clutch is an exceptionally tight fit on the clutch centre splines. Under these circumstances it is possible to distort the pressure plate permanently. Therefore the use of the correct Triumph service tool is advised.

22 Before the primary drive can be released, it is necessary to withdraw the engine sprocket. A Triumph service tool is specified; the extractor bolts thread into the two tapped holes provided. If the tool is not available, there is room for the insertion of a sprocket puller provided the two forward-mounted stator coil retaining studs are removed first. When the taper of the engine sprocket is broken, the engine sprocket and clutch assembly can be drawn off their respective shafts, together with the primary chain, and then separated.

23 Note that if the Triumph service tool is not used, the clutch centre will remain on the gearbox mainshaft. The uncaged rollers of the clutch centre bearing will be displaced as the clutch chainwheel is withdrawn; they should be collected together and placed aside for reassembly.

24 The clutch centre can now be withdrawn, using a sprocket puller to break the keyed taper joint. If there is not sufficient access to engage the legs of the puller, mainshaft end float can be gained by removing the outer cover of the gearbox and releasing the nut retaining the kickstarter ratchet on the opposite end of the mainshaft. See Chapter 2.

25 The rear portion of the primary chaincase can now be removed. It will pull off the boss around the crankcase casting quite easily.

26 Place a container that will hold at least six pints under the oil tank and remove the drain plug. Allow the oil to drain completely before disconnecting the oil pipe junction block from the rear of the right-hand crankcase. A small amount of oil will be released from this area when the joint is broken. The

engine unit is now ready to be lifted from the frame.

27 Place a wooden block under the crankcase sump to support the engine unit as the engine bolts are removed. In the case of the 6T Thunderbird model and the T110 remove the cover plate from the front engine plates, which is retained by one screw. Remove the two studs and two bolts, so that the front engine plates can be detached completely. Remove also the long stud that secures the base of the crankcase to the frame. Next, remove the two upper studs that hold the crankcase to the rear engine plates. Slacken the bottom studs and those that retain the rear engine plates to the frame. Tilt the engine unit backwards in order to release the bottom stud from the slots in the engine plates and then lift the engine unit out of the frame from the left-hand side. It is preferable to transfer the engine unit direct to the workbench, securing it firmly so that the dismantling can continue.

## 6 Dismantling the engine - general

1 Before commencing work on the engine unit, the external surfaces should be cleaned thoroughly. A motor cycle engine has very little protection from road grit and other foreign matter, which will find its way into the engine if this simple precaution is not observed. One of the proprietary cleaning compounds such as Gunk or Jizer should be used especially if the compound is allowed to work into the film of oil and grease before it is washed away. When washing down, make sure the water cannot enter the electrical system or the now exposed inlet port(s). It will, of course, be necessary to replace the rocker boxes temporarily if this method of cleansing is employed.

2 Never use undue force to remove any stubborn part, unless mention is made of this requirement. There is invariably good reason why a part is difficult to remove, often because the dismantling operation has been tackled in the wrong sequence. Dismantling will be made easier if a simple engine stand is made up that will correspond with the engine mounting points. This arrangement will permit the complete unit to be clamped rigidly to the work bench, leaving both hands free.

## 7 Dismantling the engine - removing the cylinder head, block and pistons

1 Slacken and remove the cylinder head bolts one turn at a time until the load is released. To obviate the risk of distortion, particularly if the cylinder head is of light alloy, it is advisable to slacken the bolts in diagonal sequence. There is no necessity to remove the rocker boxes at this stage and they are best left in position. If external oil drain pipes interconnect the cylinder head and the pushrod tubes, they must be removed first.

2 When all the bolts have been withdrawn, the cylinder head can be lifted off together with the cylinder head gasket. Unless damaged, in which case renewal is essential, the cylinder head gasket can be re-used after it has been annealed, as described later.

3 Lift off the pushrod cover tubes at the back and the front of the cylinder barrel. Discard the rubber seals; they must be renewed (if an oil tight engine is desired).

4 Turn the engine until both pistons are at top dead centre (TDC), then remove the eight nuts and washers around the base of the cylinder block. Before the block is lifted, place a rubber wedge between the inlet and exhaust valve tappets, to prevent them from falling from their guides as the cylinder block is raised.

5 Lift the cylinder block gently, taking care to support both pistons as they clear the cylinder bores. Slip rubber protectors over the now exposed cylinder base studs because they will damage the piston skirts if the pistons are permitted to drop free. Remove the cylinder base gasket and check that the two locating dowels are positioned correctly for subsequent reassembly.

6 Before the cylinder block is put aside, withdraw the tappets

and mark them so that they are replaced in their identical positions. Failure to observe this precaution may result in excessive tappet and cam wear and a very noisy engine.

7 Remove the circlips from each piston and push out the gudgeon pin, so that each piston can be detached from its connecting rod. It is probable the gudgeon pins will be a tight fit, in which case the pistons should be heated to expand the metal around the gudgeon pin boss. The alternative is to use Triumph service tool which will press each gudgeon pin out of position. Always support the connecting rod when any pressure is applied to either the piston or the gudgeon pin, otherwise there is risk of distortion.

8 Discard the circlips. They should never be re-used; new replacements are essential to eliminate the possibility of the circlips working loose whilst the engine is running and causing the gudgeon pin to make contact with the cylinder bore.

9 Mark each piston with pencil, inside the skirt, to ensure they are replaced in identical positions. If this precaution is not observed, a high rate of wear or oil consumption may occur.

## 8 Dismantling the engine - removing the magneto

1 Before the magneto can be removed, it is first necessary to take off the timing cover. Unscrew the ten screws around the periphery and ease the cover away. It may be necessary to give a few light taps around the edge with a rawhide mallet to start the initial separation. NEVER use anything to lever the jointing surface apart. The timing cover will come away without need to disturb the oil pressure release valve indicator.

2 Withdraw the magneto drive pinion from inside the timing chest. If the machine is fitted with an auto-advance unit, the pinion sleeve nut is of the self-extracting type. It will slacken initially, then tighten up again as it commences to draw the pinion off the tapered drive shaft of the magneto. Continue slackening until the pinion is free from the shaft. On machines fitted with manually-operated ignition advance, it is probable that a sprocket puller will be necessary to draw the pinion off the shaft, after the retaining nut has been removed. If the three nuts and washers are removed from the magneto flange on the back of the timing chest, the magneto can be lifted away as a complete unit.

## 9 Dismantling the engine - removing the distributor (alternative ignition system)

1 If the machine is of the type fitted with an alternator, dismantling is simplified. It is necessary to remove the timing cover first so that the drive pinion can be pulled off the taper of the distributor drive shaft.

2 The distributor is retained to the rear of the timing cover in the position previously occupied by the magneto on earlier models. The entire unit should be removed, complete with the mounting plate and coil assembly. Place it in a safe place whilst the dismantling continues.

3 There will not be any dynamo to remove since the alternator mounted on the end of the crankshaft supplies the electrical energy for both the ignition and lighting circuits.

## 10 Dismantling the engine - removing the oil pump and timing pinions

1 The oil pump, exposed after the timing cover has been removed, is retained in position by two conical nuts which, when removed, permit the pump to be drawn off the mounting studs. There is a paper gasket behind the pump which must be renewed each time the pump is removed.

2 To remove the timing pinions, first unscrew the nut that retains the crankshaft pinion to the end of the crankshaft and the nut retaining each camshaft pinion to its camshaft. Note that

5.22 This nut requires a box or socket spanner with thin sides

5.22a Puller may be needed to release collar from crankshaft end

7.1 Remove the external oil drain pipes

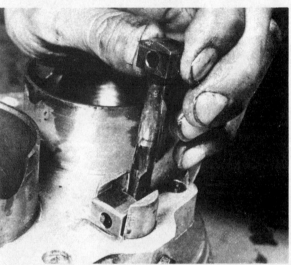
7.6 Mark tappets so that they are replaced in original order

7.7 Use screwdriver to remove circlips. Never re-use them

8.2 Remove centre nut before applying extractor (manual ignition models)

10.1 Oil pump will lift off, after removing the two retaining nuts

the camshaft pinion retaining nuts each have a LEFT HAND thread, whereas the crankshaft pinion retaining nut has a normal right hand thread. It is not necessary to remove the camshaft pinions for the crankcases to be separated, unless attention to the camshafts is required. It is preferable to leave them in-situ, with their retaining nuts in position since both pinions are very difficult to extract without the appropriate Triumph service tool.

3   The crankshaft pinion is also difficult to remove without Triumph service tool Z121. The chief danger is the risk of damage to the end of the crankshaft whilst the pinion is being drawn off; extreme care is necessary. If the Triumph service tool is not available, the pinion can be left in place until the crankcases are separated. It can then be removed by driving the crankshaft through the right hand side main bearing with a rawhide mallet. **This method is recommended only if the main bearing concerned is due for replacement.**

4   The idler pinion is easy to remove; it will pull off its centre spindle. Note how the timing pinions are marked; it will be necessary to re-align these marks in a certain manner during reassembly, to ensure the valve timing is correct. If it is desired to remove the camshaft pinions, mark the position of the key in relation to the keyway used before removing the pinions. This is IMPORTANT because each pinion has three keyways, only one of which is used.

5   Use Triumph service tool Z89 to extract the pinions. They are a tight fit on the camshafts and there is no other way to remove them without risking damage. Screw on the extractor body until it is fully engaged with the threads of the pinion boss, then screw in the extractor bolt and turn clockwise.

### 11 Dismantling the engine - separating the crankcases

1   The crankcases can now be separated by removing the various studs and bolts that hold them together. Note that there are two screws within the crankcase mouth, easily overlooked.

2   Pull the crankcases apart to release the crankshaft assembly. The camshafts will remain in the right-hand crankcase, especially if the camwheel pinions are still in place. Do not lose the small spring and disc valve that locate with the left-hand end of the inlet camshaft. They form an essential part of the crankcase breather assembly.

3   The crankshaft assembly will probably remain attached to the right-hand crankcase and can be freed by tapping the end very carefully with a rawhide mallet so that it is driven through the timing side main bearing.

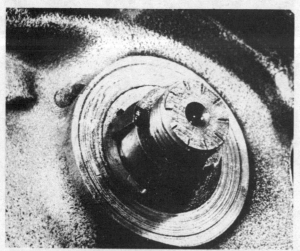

10.2 Camshaft pinion retaining nuts have a left-hand thread. Note key

10.3 The crankshaft pinion is difficult to remove without causing damage

11.1 These two screws are easily overlooked

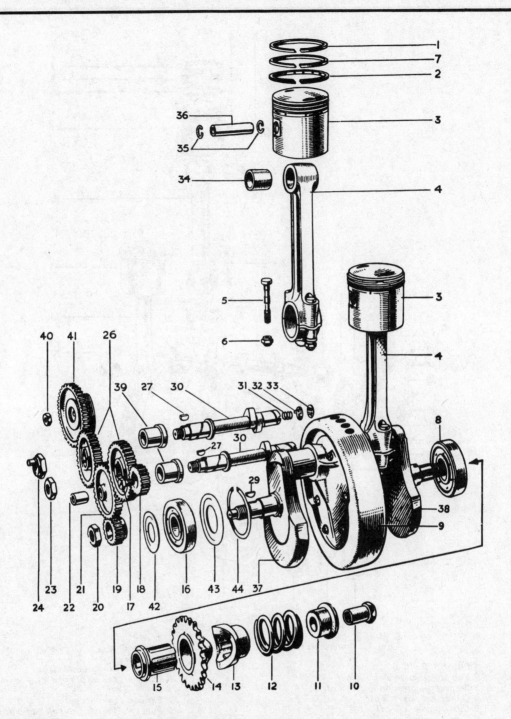

Fig. 1.2 Crankshaft (two -piece) and timing pinions, early models

1 Top compression ring - 2 off
2 Oil scraper ring - 2 off
3 Piston - 2 off
4 Connecting rod - 2 off
5 Big-end bolt - 4 off
6 Nut for big-end bolt - 4 off
7 Second compression ring - 2 off
8 Left-hand main bearing
9 Central flywheel
10 Shock absorber nut
11 Shock absorber collar
12 Shock absorber spring
13 Shock absorber slider
14 Engine sprocket

15 Shock absorber sleeve
16 Right-hand main bearing
17 Dynamo pinion screw
18 Dynamo pinion
19 Crankshaft pinion
20 Crankshaft pinion nut
21 Intermediate timing wheel pinion (idler)
22 Intermediate pinion bush
23 Exhaust camshaft pinion nut
24 Inlet camshaft pinion nut
26 Inlet and exhaust camshaft pinions
27 Key - 2 off
29 Key
30 Exhaust and inlet camshafts

31 Breather valve spring
32 Breather valve rotor
33 Breather valve disc
34 Small end bush - 2 off
35 Circlip - 4 off
36 Gudgeon pin - 2 off
37 Right-hand crank
38 Left-hand crank
39 Camshaft bushes
40 Magneto pinion nut
41 Magneto pinion (see inset for type)
42 Chip shield, right-hand
43 Right-hand bearing disc
44 Circlip

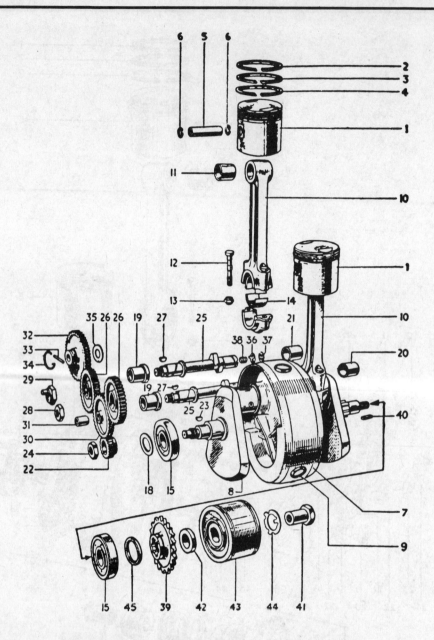

**Fig. 1.3 Crankshaft (one-piece) and timing pinions, late models**

1 Piston - 2 off
2 Top compression ring - 2 off
3 Second compression ring - 2 off
4 Oil scraper ring - 2 off
5 Gudgeon pin - 2 off
6 Circlip - 4 off
7 Flywheel
8 Crankshaft assembly
9 Bolt
10 Connecting rod - 2 off
11 Small end bush - 2 off
12 Big end bolt - 4 off
13 Nut for big end bolt - 4 off
14 Right-hand main bearing
15 Left-hand main bearing
18 Clamping washer
19 Right-hand camshaft bush - 2 off
20 Left-hand camshaft bush
21 Left-hand camshaft bush - breather type
22 Crankshaft pinion
23 Key
24 Crankshaft pinion nut

25 Inlet and exhaust camshafts
26 Camshaft timing pinions
27 Key - 2 off
28 Exhaust camshaft nut
29 Inlet camshaft nut and oil pump drive
30 Intermediate timing pinion (idler)
31 Intermediate pinion bush
32 Distributor pinion
33 Distributor pinion pin
34 Distributor pinion circlip
35 Brass washer
36 Breather valve rotor
37 Breather valve disc
38 Breather valve spring
39 Engine sprocket
40 Rotor key
41 Sprocket and rotor nut
42 Clamping washer
43 Alternator rotor
44 Lock washer
45 Oil seal

4 The left-hand crankcase will almost certainly contain the outer race of the drive side roller bearing, the inner race being on the left-hand end of the crankshaft assembly, against the crank cheek. There is no need to disturb these parts unless the bearing has to be renewed. A bearing puller (or a fine pointed wedge and a hammer) will be needed to ease the inner race off the crankshaft. To remove the outer race, heat the crankcase to expand the bearing housing. The race will drop out of position if the crankcase is brought down smartly on a block of wood, outer face upwards. If the engine is of the type that has an oil seal fitted, this will be left in position as the bearing race is removed and should be drifted out from the inside of the crankcase and renewed.

5 Note that the timing side (right-hand) main bearing is retained by a circlip, which must be removed first. There is also a chip shield, to prevent debris finding its way into any of the internal oilways (early models).

## 12 Examination and renovation - general

1 Now that the engine is stripped completely, clean all the

12.3 Now is the time to reclaim damaged or stripped threads

13.1 If there is the slightest doubt, renew the main bearings

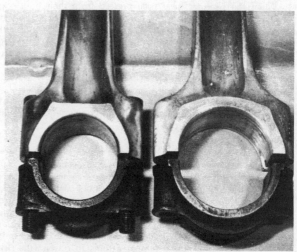

14.3 Early connecting rods do not have shell bearings

14.3a Bearings are white metalled direct

14.7 Crankshafts differ. Alternator type has no threaded end

component parts in a petrol/paraffin mix and examine them carefully for signs of wear or damage. The following sections will indicate what wear to expect and how to remove and renew the parts concerned, when renewal is necessary.

2   Examine all castings for cracks or other signs of damage. If a crack is found, and it is not possible to obtain a new component, specialist treatment will be necessary to effect a satisfactory repair.

3   Should any studs or internal threads require repair, now is the appropriate time. Care is necessary when withdrawing or replacing studs because the casting may not be too strong at certain points. Beware of overtightening; it is easy to shear a stud by overtightening giving rise to further problems, especially if the stud bottoms.

4   Where internal threads are stripped or badly worn, it is preferable to use a thread insert, rather than tap oversize. Most dealers can provide a thread reclaiming service by the use of Helicoil thread inserts. They enable the original component to be re-used.

## 13 Main bearings and oil seals - examination and renovation

1   When the bearings have been pressed from their housings, wash them in a petrol/paraffin mix to remove all traces of oil. If there is any play in the ball or roller bearings, or if they do not revolve smoothly, new replacements should be fitted. The bearings should be a tight push fit on the crankshaft assembly and a press fit in the crankcase housings. A proprietary sealant such as Locktite can be used to secure the bearings if there is evidence of a slack fit and yet they are fit for further service.

2   The crankcase oil seal should be renewed as a matter of course, whenever the engine is stripped completely. This will ensure an oiltight engine.

## 14 Crankshaft assembly - examination and renovation

1   Wash the complete crankshaft assembly with a petrol/paraffin mix to remove all surplus oil. Mark each connecting rod and cap, to ensure they are replaced in exactly the same position, then remove the cap retainer nuts so that the caps and connecting rods can be withdrawn from the crankshaft. It is best to unscrew the nuts a turn at a time, to obviate the risk of distortion.

2   Inspect the bearing surfaces for wear. Wear usually takes the form of scuffing or scoring, which will immediately be evident.

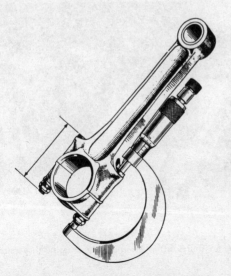

Fig. 1.4 Measuring connecting rod bolt stretch

Bearing shells are cheap to renew; it is wise to renew the shells if there is the slightest question of doubt about the originals.

3   More extensive wear will require specialist attention, either by having the crankshaft reground or by fitting a service-exchange replacement. If the crankshaft is reground, two undersizes of bearing shells can be obtained: −0.010 in and −0.020 in. It is particularly important to note that the white metal bearing shells are prefinished to give the correct diametral clearance and on no account should the bearings be scraped or the connecting rod and cap joint filed in order to achieve a satisfactory fit. If such action seems necessary, the crankshaft has not been reground to the correct tolerances. Early models have the connecting rods white metalled direct. There are no replaceable shells. In the event of damage, it will prove necessary to fit a later type crankshaft assembly.

4   Renew the connecting rod bolts and nuts as a matter of course, whenever they are disturbed.

5   It is not usually necessary to disturb the crankshaft assembly unless the lubrication system has become contaminated, in which case it may be advisable to clean out the central oil tube. Access is gained by unscrewing the retainer plug found in the right hand end of the crankshaft and then removing the flywheel bolt adjacent to the big end journal. The oil tube can be hooked out by passing a length of rod through the flywheel bolt orifice. Note that the retainer plug is retained by a centre punch mark and that it will be necessary to use an impact screwdriver, after the indentation has been drilled out.

6   Wash the oil tube in a petrol/paraffin mix and check that all the internal drillings in the crankshaft are quite free, before the tube is replaced. A jet of compressed air is best for this purpose. Make sure the retaining plug is tightened fully and centre punch the crankshaft at the screw slot, to retain the plug in position. If disturbed, the three flywheel retaining bolts should be refitted with new shakeproof washers and tightened to a torque setting of 330 - 350 lbf in (27.5 - 29.2 lbf ft/3.8 - 4.1 kgf m).

7   The earlier two-piece crankshaft assembly has separate cranks that bolt direct to the central flywheel and it is imperative that the component parts are clearly marked prior to dismantling so that there is no possibility of them being reassembled in a different order. After marking, grip the centre flywheel in a vice and slacken and remove the six nuts and bolts that retain the cranks to the flywheel. The cranks can then be parted and inspected, in conjunction with the centre oil tube.

8   When reassembling the two-piece crankshaft assembly, it is important that the centre oil tube is replaced in the correct position (see accompanying diagram) so that the cutaway engages with the spigot on the oil retainer plug. This will ensure the hole in the centre of the tube points towards the middle of the flywheel. Always fit NEW high tensile bolts and nuts and double check that all the components are in their original positions before tightening up. Centre punch the threads to prevent slackening.

## 15 Balancing the crankshaft assembly

1   If a new or re-ground crankshaft assembly is to be fitted, it should be checked for balance using a pair of 595 gram weights (Triumph part number Z120). Position the crankshaft assembly on a pair of horizontal knife edges, after checking with a spirit level that the knife edges are truly horizontal. Mark the lowest part of the assembly with chalk and rotate it through 90°. If the assembly returns to the original position, drill at the chalk mark on the centre line of the flywheel, using a ¼ in. diameter drill. Drill a little at a time, constantly stopping to check and never drill deeper than ½ inch. If more weight has to be removed, start a new hole with the centres ¾ inch distant. When the balance is correct, the assembly should rest in any position and if disturbed, settle without returning to any one set position. Make sure the assembly is cleaned thoroughly before reassembly, to remove traces of swarf and dirt.

2   The 595 gram balance weights are used to compensate for

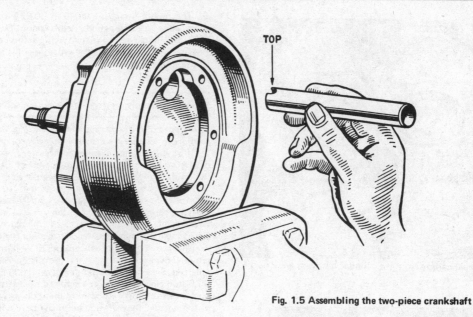

TOP

**Fig. 1.5 Assembling the two-piece crankshaft**

the weight of the pistons, gudgeon pins and small end bushes, etc. One should be suspended from each connecting rod, during the balancing operation.

## 16 Camshaft and timing pinion bushes - examination and renovation

1   It is unlikely that the camshaft and timing pinion bushes will require attention unless the machine has covered a high mileage. The normal rate of wear is low. Bushes in the right hand crankcase can be removed by heating the crankcase to expand the surrounding metal and driving them out from the outside, using a two-diameter drift of the correct size. Fit the new bushes whilst the crankcase is still hot and make sure they are correctly aligned so that any oil feed holes register with those of the crankcase.

2   The blind bushes in the left hand crankcase are more difficult to remove. The recommended technique is to tap the bushes with a Whitworth thread and screw home a bolt of matching thread. If the crankcase is now heated, the bolt head can be gripped in a vice and the crankcase driven off the bush with a rawhide mallet. When the inlet camshaft bush is replaced, care must be taken to engage the peg with the breather porting disc that lies behind the bush.

3   The camshaft bushes are machined from sintered bronze and only the smallest amount of metal will need to be removed after they are pressed into position. The intermediate (idler) timing gear bush is machined from phosphor bronze.

## 17 Camshafts, tappet followers and timing pinions - examination and renovation

1   Examine each camshaft, checking for wear on the cam form, which is usually evident on the opening flank and on the lobe. If the cams are grooved, or if there are scuff or score marks that cannot be removed by light dressing with an oilstone, the camshaft concerned should be renewed.

2   When extensive wear has necessitated the renewal of a camshaft, the camshaft and tappet followers should be renewed at the same time. It is false economy to use the existing camshaft followers with a new camshaft since they will promote a more rapid rate of wear.

3   At some stage, the question will inevitably arise whether to fit camshafts to improve performance, such as the E3134 type.

Provided it is acceptable that the machine will be less flexible at low speeds and the rate of petrol consumption may rise, no problems will be encountered. Make sure that the matching cam followers (tappets) are fitted at the same time. Apart from being bad engineering practice, the running together of old and new parts will result in a high rate of wear, whilst the use of unsuitable tappets may give unexpected variations in the valve timing.

4   Check each timing pinion for worn or broken teeth. Damage is most likely to occur if some engine component has failed during service and particles of metal have circulated with the lubrication system. Excessive backlash in the pinions will lead to noisy timing gear.

## 18 Cylinder block - examination and renovation

1   There will probably be a lip at the uppermost end of each cylinder bore that denotes the limit of travel of the top piston ring. The depth of the lip will give some indication of the amount of bore wear that has taken place, even though the amount of wear is not evenly distributed.

2   Remove the rings from the pistons, taking great care as they are brittle and easily broken. Most wear occurs within the top half of the bore, so the pistons should be inserted and the clearance between the skirt and the cylinder wall measured. If measurement by feeler gauge shows the clearance is 0.005 in greater or more than the figure quoted in the Specifications Section, the cylinder is due for a rebore. Oversize pistons are supplied in three sizes: + 0.010 in, +0.020 in and + 0.040 in; the cylinder should be rebored to suit.

3   Give the cylinder block a close visual inspection. If the surface of either of the bores is scored or grooved, indicative of a previous engine seizure or a displaced circlip and gudgeon pin, a rebore is essential. Compression loss will have a very marked effect on performance.

4   Check that the outside of the cylinder block is clean and free from road dirt. Use a wire brush on the cooling fins if they are obstructed in any way. The application of matt cylinder black will help improve the heat radiation, if the block is of cast iron.

5   Check that the base flange is not cracked or damaged. If the engine has been overstressed, one of the first parts to fail is the base of the cylinder barrel, either at the holding down points or around the base of each bore. If a crack is found, the cylinder barrel should be renewed.

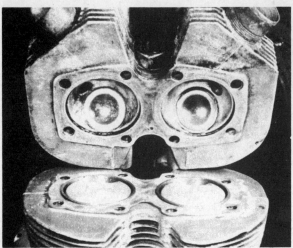
21.1 Remove all carbon from the cylinder head and grind in the valves

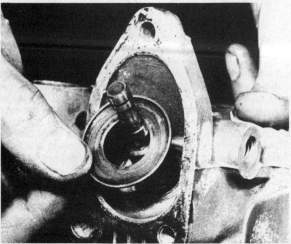
21.8 Don't omit valve spring seat

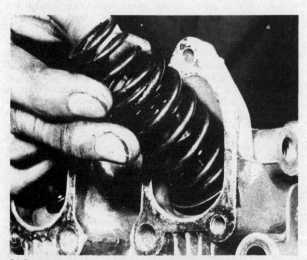
21.8a Check length of valve springs before reassembly

6   The rebore limit is + 0.040 in. (0.020 in, all alloy engines). Above this size, the cylinder walls cannot be considered to have sufficient thickness consistent with safety and reliability. Re-sleeving or a service exchange replacement is the only practicable solution to the problem.

### 19 Pistons and piston rings - examination and renovation

1   Attention to the pistons and piston rings can be overlooked if a rebore is necessary, since new replacements will be fitted.
2   If a rebore is not considered necessary, examine each piston closely. Reject pistons that are scored or badly discoloured as the result of exhaust gases by-passing the rings.
3   Remove all carbon from the piston crowns, using a blunt scraper which will not damage the surface of the piston. Clean away all carbon deposits from the valve cutaways and finish off with metal polish so that a clean, shining surface is achieved. Carbon will not adhere so readily to a polished surface.
4   Check that the gudgeon pin boses are not worn or the circlip grooves damaged. Check that the piston ring grooves are not enlarged. Side float should not exceed 0.003 in.
5   Piston ring wear can be measured by inserting the rings in the bore from the top and pushing them down with the base of the piston so that they are square in the bore and about 1½ inches down. If the end gap exceeds 0.014 in, all rings, renewal is necessary. The exception is the oil scraper ring of the 500 cc models, the end gap of which should not exceed 0.011 in.
6   Check that there is no build up of carbon on the inside surface of the rings or in the grooves of the pistons. Any buildup should be remvoed by careful scraping. Piston rings must be fitted with the land marked 'TOP' uppermost, if taper faced.
7   The piston crowns will show whether the engine has been rebored on some previous occasion. All oversize pistons have the rebore size stamped on the crown. This information is essential when ordering replacement piston rings.

### 20 Small end bearings - examination and renovation

1   The amount of wear in the small end bushes can be ascertained by the fit of the gudgeon pins. The pin should be a good sliding fit in each case, without evidence of any play.
2   Renewal can be effected by using a simple drawbolt arrangement (as illustrated) whereby the new bush is used to press the old bush out of location. It is essential to ensure the oilway in the bush locates with the oilways in the connecting rod, otherwise the bearing will run dry and rapid wear will occur.
3   After the bushes have been fitted, they will have to be reamed out to the correct size. Cover the mouth of the crankcase with rag to prevent metallic particles from dropping in.

### 21 Cylinder head and valves - dismantling, examination and renovation

1   It is best to remove all carbon deposits from the combustion chambers, before removing the valves for grinding-in. Use a blunt-ended scraper so that the surface of the combustion chambers is not damaged and finish off with metal polish to achieve a smooth, shiny surface.
2   Before the valves can be removed, it is necessary to obtain a valve spring compressor of the correct size. This is necessary to compress each set of valve springs in turn, so that the split collets can be removed from the valve cap and the valve and valve spring assembly released. Keep each set of parts separate; there is no fear of inadvertently interchanging the valves because the heads are marked 'IN' or 'EX'.
3   Before giving the valves and valve seats further attention, check the clearance between each valve stem and the valve guide in which it operates. Some play is essential in view of the high

temperatures involved, but if the play appears excessive, the valve guides must be renewed.

4   To remove the old valve guides, heat the cylinder head and drive them out of position with a double diameter drift of the correct size. Replace the new guides, whilst the cylinder head is still warm. Note that when bronze valve guides are fitted, the two SHORT guides are fitted in the inlet position. Conversely, when cast iron guides are fitted, the two LONG guides are fitted in the inlet position.

5   Grinding in will be necessary, irrespective of whether new valve guides have been fitted. This action is necessary to remove the indentations in the valve seats caused under normal running conditions. It is also necessary when new valve guides have been fitted, in order to re-align the face of each valve with its seating.

6   Valve grinding is a simple task. Commence by smearing a trace of fine valve grinding compound (carborundum paste) on the valve seat and apply a suction tool to the head of the valve. Oil the valve stem and insert the valve in the guide so that the two surfaces to be ground in make contact with one another. With a semi-rotary motion, grind in the valve head to the seat, using a backward and forward action. Lift the valve occasionally so that the grinding compound is distributed evenly. Repeat the operation until a ring of light grey matt finish is obtained on both valve and seat. This denotes the grinding operation is complete. Before passing to the next valve, make sure that all traces of compound have been removed from both the valve and its seat and that none has entered the valve guide. If this precaution is not observed, rapid wear will take place due to the abrasive nature of the carborundum base.

7   When deeper pit marks are encountered, or if the fitting of a new valve guide makes it difficult to obtain a satisfactory seating, it will be necessary to use a valve seat cutter set to an angle of 45º and a valve refacing machine. This course of action should be resorted to, only in an extreme case, because there is risk of pocketing the valve and reducing performance. If the valve itself is badly pitted, fit a replacement.

8   Before reassembling the cylinder head, make sure that the split collets and the taper with which they locate on each valve are in good condition. If the collets work loose whilst the engine is running, a valve will drop and cause extensive engine damage. Check the free length of the valve springs with the Specifications Section and renew any that have taken a permanent set.

9   Reassemble by reversing the procedure used for dismantling the valve gear. Do not neglect to oil each valve stem before the valve is replaced in the guide.

10   Before setting aside the cylinder head for reassembly, make sure that cooling fins are clean and free from road dirt. Check that no cracks are evident, especially in the vicinity of the holes through which the holding down studs and bolts pass, and near the spark plug threads.

11   Finally, make sure that the cylinder head flange is completely free from distortion at the joint it makes with the cylinder barrel. An aluminium alloy cylinder head will distort with comparative ease if it is tightened unevenly and may lead to a spate of blowing cylinder head gaskets. If the amount of distortion is not too great, flatness can be restored by careful rubbing down on a sheet of fine emery cloth wrapped around a sheet of plate glass. Otherwise it may be necessary to have the cylinder head flange refaced by a machining operation.

---

**22 Tappets and tappet guide blocks - examination and renovation**

1   Mention has not been made of the tappets or tappet guide blocks, which seldom require attention. The amount of wear within the tappet blocks can be ascertained by rocking the tappet whilst it is within the tappet block. It should be a good sliding fit, with very little sideways movement.

2   To remove and replace the tappet blocks, first remove the locking screw. The block can then be drifted out of position, preferably by using the Triumph service tool.

3   To replace the tappet blocks, first grease the outer surface, and align the hole for the locking screw before driving the block

22.1 Tappet blocks rarely require attention

22.2 Remove screw before tappet block is driven out

22.2a Drift out of position with great care

23.4 Inlet and exhaust rocker boxes are separate

24.1 Check return pipe to scavenge pump is unobstructed

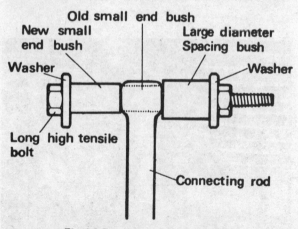

Fig. 1.6 Renewing the small end bush

back into position, with the shoulder flush against the flange. The Triumph service tool is again recommended to simplify this operation.

4    As mentioned in Section 17.2, the tappet followers should be renewed if they show signs of wear. Make sure that the replacements are fitted correctly, with the machined cut-away in

the stem facing the outside of the tappet guide block. **If fitted incorrectly, the tappets will not be lubricated.** It is also important to note that the inlet and exhaust tappets and the tappet guide blocks must not be interchanged.

## 23 Pushrods, rocker spindles and rocker arms - examination and renovation

1    Check the pushrods for straightness by rolling them on a sheet of glass. If any are bent, they should be renewed since it is not easy to effect a satisfactory repair.
2    Check the end pieces to ensure that they are a tight fit on the light alloy tubes. If the end pieces work loose, the pushrod must be renewed. It is unlikely that the end pieces will show signs of wear at the point where they make contact with the rocker arms and the tappet followers, unless the machine has covered a very high mileage. Wear usually takes the form of chipping or breaking through the hardening, which will necessitate renewal.
3    Examine the tips of the rocker arms. If wear is evident, both the valve clearance adjusters and the ball pins should be renewed. The latter should be pressed into place with the drilled flat towards the rocker spindle.
4    If it is necessary to renew the rocker spindles, they can be driven out of the rocker box housing by means of a soft metal drift. Reassembly is best accomplished by using either the Triumph service tool or a 7/16 in. bolt six inches long with a taper ground on the end. Note that each spindle has a plain washer with a small diameter bore that acts as a thrust washer and is assembled last against the inner right hand face of the rocker box. The oil feed to the rocker spindles must be on the right hand side of the machine. Before fitting the spindles, check that the oilways are clean and unobstructed, preferably by using a jet of compressed air. Lubricate the spindle thoroughly before it is inserted.
5    There is no adjustment for end float. This function is performed by the spring washers, fitted between each end of the rocker arm and the rocker box cover.

## 24 Engine reassembly - general

1    Before the engine is reassembled, all the various components must be cleaned thoroughly so that all traces of old oil, sludge, dirt and gaskets etc are removed. Wipe each part with clean, dry lint-free rag to make sure there is nothing to block the internal oilways of the engine during reassembly.
2    Make sure that all traces of old gaskets have been removed and that the mating surfaces are clean and undamaged. One of the best ways to remove old gasket cement is apply a rag soaked in methylated spirit. This acts as a solvent and will ensure the cement is removed without resort to scraping and subsequent risk of damage.
3    Gather together all the necessary tools and have available an oil can filled with clean engine oil. Make sure the new gaskets and oil seals are to hand; nothing is more infuriating than having to stop in the middle of a reassembly sequence because a vital gasket or replacement has been overlooked.
4    Make sure that the reassembly area is clean and that there is adequate working space. Refer to the torque and clearance settings whenever they are given. Many of the smaller bolts are easily sheared if they are overtightened. Always use the correct size screwdriver.

## 25 Engine reassembly - rebuilding the crankshaft

1    Refit the connecting rods to the crankshaft assembly in their original positions, using the marks made during the dismantling operation as a guide. Make sure that the shell bearings are located correctly, then replace the end caps, the retaining nuts

and bolts. Tighten each nut evenly until the connecting rods and their end caps seat correctly, then tighten the nuts with a torque wrench to a setting of 320 - 330 lbf in (26.7 - 27.5 lbf ft/3.7 - 3.8 kgf m). *Note this only applies to later models with connecting rods which have separate bearing shells.* A more accurate method of tightening the nuts is to use a micrometer to measure the amount of bolt stretch (see Fig. 1.4). Measure the bolt lengths before fitting, then tighten the nuts evenly until the bolts have stretched by 0.004 - 0.005 in (0.1016 - 0.1270 mm). This latter method is preferable to the use of a torque wrench because it ensures that the bolts are accurately tensioned with the minimum risk of overstress. On all earlier models with white metal lined connecting rods, tighten the nuts evenly until the bolts have stretched by 0.007 - 0.008 in (0.1778 - 0.2032 mm). On all models check that the connecting rods are free to rotate, with no discernible free play.

2   Apply a pressure oil can to the drilling at the right hand end of the crankshaft and pump until oil is expelled from both big ends. This is essential, to ensure that the oil passages are free from obstruction and full of oil.

26.4 Rotary breather fits within left-hand end of inlet camshaft housing

26.4a Oil all bearings before lowering camshaft assembly into position

27.1 Inexpert removal of pinion will cause this end of crankshaft to bell out

## 26 Engine reassembly - reassembling the crankcases

1   If the main bearings have been removed, replace the ball journal bearing in the right hand crankcase and the outer race of the roller bearing in the left hand crankcase. It is advisable to heat the respective crankcase beforehand so that the bearings will drop into place without difficulty.

2   The inner race of the roller bearing should be driven on to the left hand side of the crankshaft assembly until it is hard against the shoulder of the crankshaft.

3   Any shims that have been fitted to limit the end float of the crankshaft assembly must be fitted BEHIND the main bearings in the crankcases. They should be of equal thickness on either side, so that the crankshaft assembly is centrally disposed within the crankcase.

4   Mount the left hand crankcase on two blocks of wood so that there will be sufficient clearance for the end of the crankshaft to project downwards without touching the bench. Lubricate the main bearings and the camshaft bushes and place the rotary breather valve and spring in the inlet camshaft bush. If the camshafts are not fitted to the right hand crankcase, position them in their respective bushes taking care that the slot in the end of the inlet camshaft engages with the projection of the rotary breather valve.

5   Lower the crankshaft assembly into position and give it a sharp tap to ensure that the inner race of the roller bearing is fully engaged with the outer race.

6   Coat the jointing face of the right hand crankcase with gasket cement and lower it into position, after checking that the connecting rods are centrally disposed. Do not omit to replace the chip shield and circlip in front of the timing side main bearing (early models). If the camshafts are still attached to the right hand crankcase it will be necessary to rotate the inlet one until it engages with the rotary breather valve before the crankcases will meet. Push both crankcases together so that they mate correctly all round and check that the crankshaft and the camshafts will revolve quite freely before the securing bolts and studs are replaced and tightened by hand.

7   Check that the cylinder barrel junction of the crankcases is level and if necessary adjust by light tapping. When a level surface is achieved and the crankshaft and camshafts are free to revolve, the retaining nuts and bolts can be tightened evenly and securely.

## 27 Engine reassembly - replacing the timing pinions, oil pump and timing cover

1   Place the clamping washer, chamfered side inwards, on the end of the crankshaft. Replace the Woodruff key in the right

27.3 Timing marks make realignment of pinions easy

27.5 Use of gasket cement will block up oilways

27.6 Temporarily replace timing cover

hand end of the crankshaft and slide the crankshaft pinion into position so that the keyway locates with the key. Drive the pinion on to the end of the crankshaft until it is fully home.

2   If the camshaft pinions have been removed, position the keys in the respective camshafts which will have been marked to ensure that the correct keyways are used. Replace the pinions, again ensuring that the previously marked keyway is the one aligned. Triumph service tool Z89 can be used for replacing the pinions, in conjunction with replacer adaptor Z144. Alternatively, the pinions can be drifted on to their respective shafts using a hollow tube. Replace the pinion locknuts, noting that each has a LEFT HAND THREAD. The inlet camshaft nut has the projecting peg for the oil pump drive.

3   Replace the idler pinion on its shaft and align the pinions so that the timing marks correspond EXACTLY as shown in the accompanying diagram. If these marks align correctly, the valve timing is correct. It is best to remove and replace the idler pinion when aligning the timing marks because the hunting tooth principle used means the marks will coincide only once every 94th revolution.

4   Replace the crankshaft pinion securing nut which may otherwise obscure the timing mark of the pinion. This has a normal right hand thread and should be tightened fully.

5   The oil pump can now be replaced as a complete unit. Fit a new gasket between the oil pump body and the oilways in the timing case, making sure that the holes in the gasket align correctly. DO NOT USE GASKET CEMENT AT THIS JOINT. Thread the pump body over the two retaining studs and engage the driving peg of the inlet camshaft pinion retaining nut with the hole in the sliding block at the top of the pump. Make sure the pump body lies flat against the crankcase without any strain and replace the two conical retaining nuts. The conical portion of the nuts should face inwards; tighten the nuts securely.

6   It is preferable to temporarily replace the timing cover at this stage, even if it has to be removed after the magneto is fitted in order to time the ignition. If it is retained by just two or three screws, this will prevent dust or other foreign matter entering the timing chest whilst the engine rebuild continues.

## 28 Engine reassembly - replacing the pistons and cylinder block

1   Stand the engine upright and arrange the crankshaft so that both connecting rods are fully extended. Add about 1/6 pint of engine oil to the crankcase, pad the mouth of the crankcase with clean rag to prevent any displaced parts from falling in; commence engine reassembly.

2   Oil both gudgeon pins and small end bushes. Check that the oilway in each small end bush lines up with the oil hole in the connecting rod.

3   Warm both pistons in hot water to make the insertion of the gudgeon pin easier. Make sure the pistons are replaced in their original positions, using the markings made previously within the skirt. Do not reverse them, back to front.

4   After inserting the gudgeon pins and circlips, check that each circlip is located positively in its groove. A displaced circlip will cause severe engine damage if it works loose whilst the engine is running.

5   Always fit new circlips. Never re-use the originals.

6   Fit a new cylinder base gasket (NO gasket cement) and check to ensure that the oil holes line up with the drillings in the crankcase flange. Failure to observe this precaution will result in the tappets being starved. Check that both dowels in the left hand crankcase are positioned correctly.

7   Replace the piston rings. The compression rings are tapered and it is important to ensure the land marked 'top' is fitted uppermost in each case. When the rings are fitted, space them out so that the end gaps do not coincide and fit a pair of piston ring clamps. Place two pieces of wood below the skirt of each piston and rotate the crankshaft so that the pistons seat on these 'stops' to hold them steady.

8   Oil the tappet followers in the cylinder barrel and retain

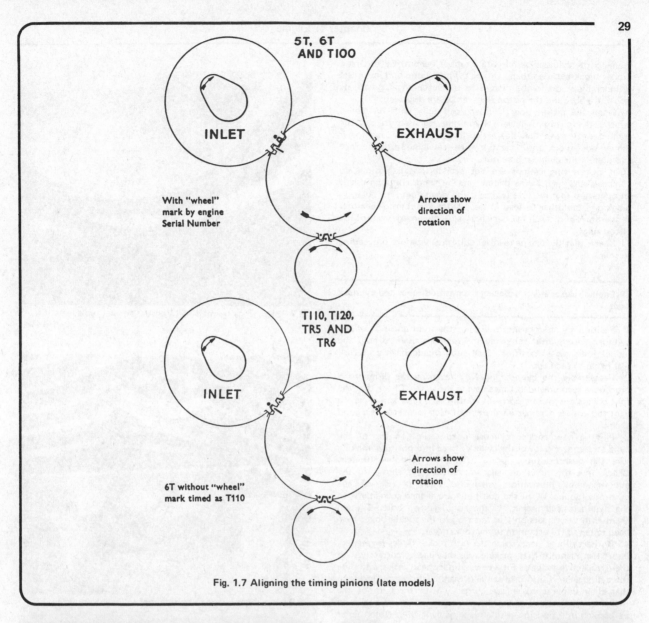

5T, 6T
AND T100

INLET

EXHAUST

With "wheel"
mark by engine
Serial Number

Arrows show
direction of
rotation

T110, T120,
TR5 AND
TR6

INLET

EXHAUST

6T without "wheel"
mark timed as T110

Arrows show
direction of
rotation

Fig. 1.7 Aligning the timing pinions (late models)

28.3 Insert gudgeon pins after warming pistons, especially if new

28.6 Use new cylinder base gasket and ....

them to the cylinder barrel with a rubber band or by forcing a rubber block between them. Oil the cylinder bores and lower the cylinder block on to the pistons so that each engages with its respective bore and the piston ring clamps are displaced.

9  When the piston rings have engaged fully with the bores, remove the piston ring clamps and the rag used to pad the crankcase. Remove the 'steadies' below each piston and lower the cylinder block until it seats over the retaining studs. Replace and tighten the cylinder base nuts.

10  If piston ring clamps are not available, it is possible to compress and feed in the piston rings by hand, although this is an operation that requires two persons - one to hold the cylinder block and the other to feed in the pistons and rings. A chamfer at the bottom of each cylinder bore aids the engagement of the piston rings.

11  Check that the engine revolves quite freely before proceeding further.

## 29 Engine reassembly - replacing the pushrod tubes and cylinder head

1  Position the two pushrod tubes at the front and rear of the cylinder barrel, after removing the retainers used to hold the tappet followers in position. Each tube should have a new oil seal fitted to each end.

2  Reassemble the cylinder head by reversing the dismantling procedure recommended for removing the valves. Check that the split collets are located correctly within each valve spring cap - a light tap with a hammer on the end of each valve stem is a good check.

3  Place a new copper cylinder head gasket on top of the cylinder barrel and lower the cylinder head into position, making sure the pushrod tubes locate with their respective tunnels. Check that the tubes are aligned correctly, as shown in the accompanying illustration, otherwise there is risk of the pushrods fouling. When the bolt holes are aligned correctly with the cylinder head gasket, fit the holding-down bolts. Tighten them with the fingers only at this stage if the rocker boxes have been removed for attention to the rocker gear.

4  If the cylinder head complete with rocker boxes is refitted, insert the pushrods first, making sure they engage correctly with their respective tappets. Fit a new cylinder head gasket and slide the cylinder head into position and manipulate the pushrods so that their upper ends engage correctly with the ends of the rocker arms. Removal of the inspection caps will permit them to be hooked into position with a piece of bent wire. Note that at

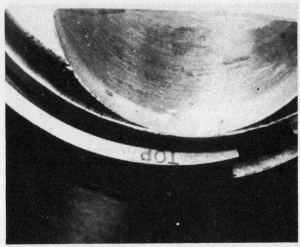

28.7 Piston rings are marked to ensure correct replacement

28.8 Use rubber bands to hold tappet followers in place

28.6a .... don't forget dowels

28.8a Clamps make fitting of cylinder block easy

28.8b Note 'steady' under pistons

29.1 Always use new seals at top and bottom of push rod tubes

29.1a Make sure tubes seat correctly and are not inverted

29.3 Cylinder head gasket can be reclaimed by annealing

29.3a Push rods must engage correctly with tappet followers

30.1 It is convenient to fit magneto at this stage

least one valve will be open whatever the crankshaft position. Fit the cylinder head bolts, making sure they pass through the cylinder head gasket correctly and tighten them in the order shown in Fig. 1.8 to a torque setting of 18 lbf ft (2.5 kgf m). This will obviate the risk of distortion, particularly with a light alloy cylinder head.

## 30 Engine reassembly - refitting the magneto and timing the ignition

1   The magneto bolts direct to the back of the timing chest. It has a flange mounting and is retained by three nuts and washers. Before the magneto is fitted, position a new gasket over the three studs that project from the timing chest, to maintain an oiltight joint. Make sure the nuts are tight.

2   Detach the timing cover that has been temporarily refitted. If the machine is fitted with automatic ignition advance, it will be necessary to wedge the auto-advance mechanism in the fully extended position whilst the pinion is replaced on the end of the magneto drive shaft. Before the pinion retaining sleeve nut is tightened, the engine and magneto contact breaker assembly must be positioned correctly, with the points about to separate.

3   Turn the engine over in a forwards direction until the piston in the right-hand cylinder is at top dead centre on the compression stroke. In this position, the inlet and the exhaust valves will be fully closed. This position should be obtained with the spark plugs removed, if necessary turning the engine by a spanner applied to the nut on the right-hand end of the crankshaft. A timing stick inserted through the plug hole will serve to indicate when the piston is at the top of the stroke; make sure the timing stick cannot fall in if the engine is turned too rapidly. Mark the stick against a fixed reference point.

4   Turn the engine backwards beyond the desired amount of ignition advance, then turn it slowly in a forwards direction until the piston is in the desired position (see Specifications, Chapter 5). Check that the magneto position has not changed, then drive the magneto pinion home on the magneto drive shaft taper, using a box or socket spanner against the face of the pinion and administering a sharp tap with a hammer to the end. Replace the retaining nut and washer and tighten it (manual advance models).

5   On models fitted with the auto-advance unit, the centre sleeve nut is captive with the unit itself and after wedging the unit in the fully advanced position and checking the magneto is positioned correctly with the points about to separate, the nut must be turned so that the unit is gradually drawn on to the drive shaft taper. This takes a little skill, since the magneto armature must not be permitted to turn even a fraction until the pinion is correctly in position. Make sure it is tight.

6   After timing the ignition, always re-check and if the setting is not correct, start all over again. Engine performance depends very much on the accuracy with which the timing is set. Good enough is never satisfactory, since even minor errors will have a surprising effect on performance. For really accurate timing, the use of a timing disc is advised.

7   It is imperative that the contact breaker gap is first checked and if necessary, re-set, before the ignition timing procedure is followed. If the gap happens to be incorrect, adjustments made afterwards will affect the accuracy of the ignition timing setting. During the timing procedure, a piece of cigarette paper between the points will give the best indication of when they are about to separate. It will be immediately obvious when the grip on the paper is released, if it is pulled gently.

8   Replace the timing cover and the timing cover screws. Use a new gasket at the joint face, and a light smear of gasket cement.

## 31 Engine reassembly - refitting the distributor and timing the ignition

*Alternator models only*

1   Assemble the distributor (complete with clamping lever) to the adaptor and tighten the retaining bolt. Fit the adaptor to the rear of the timing cover, with the clamp nut and bolt towards the crankcase and with the slotted head pointing downwards. Fit but do not tighten the lower retaining nut.

2   Clamp the coil to the bracket and ensure both bolts are tight. Assemble the coil and bracket on to the upper distributor adaptor studs and fit the nuts. Tighten all three nuts. The distributor clamp nut should be left sufficiently slack to permit the distributor body to rotate for the final positioning.

3   The position of the distributor should be arranged to give the most convenient access for timing the ignition and any subsequent maintenance. This is achieved by rotating the distributor body so that the contact breaker points are in the 11 o'clock position, when viewed from the left-hand side of the machine. Tighten the lever clamp bolt with the distributor in this position.

4   With the right-hand piston at top dead centre on the compression stroke, rotate the rotor arm clockwise until the points are just beginning to separate. The rotor arm should then be pointing towards the rear of the machine. Hold the rotor arm in this position, and slide the thrust washer and drive pinion on to the distributor shaft taper so that the hole in the boss of the pinion aligns with the hole in the drive shaft. Mesh the pinion in the nearest position of the pinion that mates with it and slide the locking pin through the pinion boss and shaft, retaining it in position by means of the circlip that locates with the groove in the pinion boss. The distributor is now in approximately the correct position for the ignition to be timed with accuracy.

5   To time the ignition, follow the procedure given in the preceding Section, paragraphs 3 and 4. In this case, the drive pinion is already positioned on the shaft. Rotate the distributor body to obtain the precise point at which the contacts are about to separate and lock the distributor in this position by tightening the clamp bolt. As an aid to the correct setting, the ignition should be switched on during the latter stages of the operation. If the ammeter is watched, the discharge reading will cease and the needle return to zero, immediately the contact breaker points separate. Do not omit to recheck the setting when everything has been tightened up.

6   Refit the timing cover and timing cover screws, using a new gasket and a smear of gasket cement.

## 32 Engine reassembly - refitting the rocker boxes

1   It may be possible to refit the rocker boxes after the engine has been installed in the frame but this depends on the model being worked on; in some cases insufficient clearance exists between the frame and cylinder head to do so. However, as more access is available for the correct location of the push rods, it is probably better to fit the rocker boxes at this stage. Fit new base gaskets with only a slight smear of gasket cement. Refit the cylinder head bolts and tighten them in the sequence shown in Fig. 1.8 to a torque setting of 18 lbf ft (2.5 kgf m).

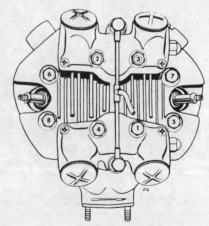

Fig. 1.8. Cylinder head bolt tightening sequence

32.2 Use new washers at all oil union joints

32.2a Take care to prevent thin pipes from 'necking'

33.2 Rear engine plate mountings

33.2a Note distance piece on footrest rod

34.1 Rear half of chaincase fits around this boss

34.1a Note sliding oil seal behind clutch and clutch pushrod

34.1b Shock absorber nut must be tightened fully

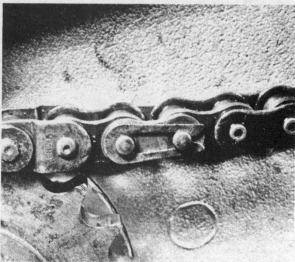

34.1c If chain has spring link, closed end faces direction of travel

35.2 Adjust clearances as shown, using square end adjuster

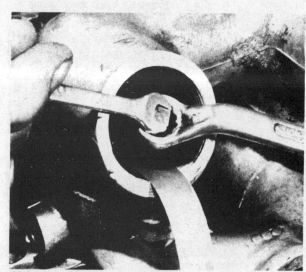

35.3 Always re-check after adjustments are made

36.1 Reattach rocker feed pipes. Note how one pipe has twisted

36.2 Refit carburettor to induction flange

2 The oil drain pipes from the cylinder head to the push rod tubes can be reconnected at the same time. Use new sealing washers at the back and front of each union joint and tighten the bolts fully. Restrain the unions from turning as the bolts are tightened, otherwise there is risk of the small diameter oil pipes 'necking' and cutting off the oil flow.

3 Check that all the rocker box bolts are tight, and especially the nuts on the underside.

## 33 Replacing the engine unit in the frame

1 When the engine unit is ready for reinstallation in the frame, check that the rear lower stud is in position, with the nuts loosely threaded on to the ends. Make sure the footrest spindle locates the distance piece between the two rear engine plates, through which it passes.

2 Lift the engine into approximately the correct position, having a stout wooden box or some similar support below the frame, on which to rest it. It is advisable to have a second person available during this task, so that one can steady the frame whilst the other lifts the engine unit into position.

3 Insert the long stud which secures the crankcase to the lug on the frame tubes. Refit the front engine plates with the various studs and nuts, then tighten all the nuts and bolts. On 6T and T110 models, replace the additional front engine place cover and tighten the screw that retains it. All the engine plate bolts must be tightened fully. If due attention is not paid to this point, engine vibration will be experienced.

4 Refit the cylinder head torque stays and tighten them fully, also the clamp around the front down tube of the frame.

5 Replace the right-hand footrest and adaptor on the footrest rod.

## 34 Engine reassembly - replacing the clutch, engine sprocket and primary chain

### Separate magneto and dynamo models

1 With these models, the procedure is somewhat simplified because the primary chain is not of the endless type. Fit the rear half of the chaincase, then the engine sprocket and shock absorber assembly, followed by the clutch. The clutch centre nut and engine sprocket/shock absorber nut must be tightened securely. The latter is best tightened last of all, so that the engine can be locked in position by engaging top gear and applying the rear brake. Do not omit to bend the tab washer over the clutch centre nut, before the clutch is assembled.

### Alternator models

2 The primary chain fitted to the alternator models is of the endless type. No spring link is fitted; the chain must be fitted together with the engine sprocket and clutch, at one and the same time. Before commencing assembly, slacken off the gearbox chain tensioner by turning in a clockwise direction.

3 Replace the back half of the primary chaincase. It is a push fit over the large diameter boss of the drive-side crankcase.

4 Replace the clutch centre complete with roller bearings and clutch chainwheel after the Woodruff key has been inserted in the gearbox mainshaft, together with the engine sprocket and chain loop. The engine sprocket must be positioned so that the taper ground boss is closest to the crankshaft main bearing; the sprocket is a light drive fit on the crankshaft splines. Tap the clutch centre to lock it on to the gearbox mainshaft.

5 Fit the clutch inner drum over the splines of the clutch centre and replace the tab washer nut that retains the assembly in position. Scotch the clutch inner drum and tighten the centre nut securely.

6 Replace the clutch plates, commencing with an inserted plate and then following with alternate plain and inserted plates. Fit the domes pressure plate, after the clutch pushrod has been inserted in the centre of the hollow mainshaft, and replace the thimbles, clutch springs and clutch adjusting nuts. Tighten the adjusting nuts evenly until the thread of the stud projecting from the clutch inner drum is flush with the end.

7 Replace the two forward-mounted studs which were removed to make the withdrawal of the engine sprocket easier and the sleeve nut in the chaincase rear, through which the leads from the generator stator coils pass. Fit the distance piece, located between the engine sprocket and the generator rotor, then position the key in the left hand end of the crankshaft and slide the generator rotor into position after aligning the keyway. Refit the tab washer and the centre retaining nut, tightening the nut securely. Bend the tab washer to lock the nut.

8 Slip the distance pieces over the stator coil retaining studs and position the stator coil assembly so that the lead connecting the coils is at the top. Thread the lead through the sleeve nut in the rear of the chaincase, until it emerges from the back. Replace the split plug that seals the sleeve nut orifice and the rubber cap over the sleeve nut extension. Check that there is no possibility of the lead fouling the primary chain. The lead must emerge from the outer portion of the stator coil assembly and be held in the clip-by the rear of the engine sprocket.

9 Replace the stator plate retaining nuts and washers, then check that the rotor does not foul the stator plate assembly when the engine is turned over. There should be a minimum clearance of 0.008 in between each of the stator coil pole pieces and the rotor.

10 Retension the primary chain by turning the gearbox adjuster in an anticlockwise direction to draw the gearbox backwards. The tension is correct when there is approximately ½ inch up and down play in the middle of the lower run.

11 Refit the front cover of the chaincase, using a new gasket at the joint, with a liberal smear of gasket cement. Replace and tighten the twelve screws around the periphery, then position the left-hand footrest and tighten the securing nut on the footrest bar.

## 35 Adjusting the valve clearances

1 Before refitting the inspection caps or finned covers to the rocker boxes, it is necessary to check, and if necessary reset, the valve clearances. To check the left hand valve clearance, turn the engine over until the right hand exhaust valve is fully open. This will ensure that the left hand tappet is resting on the base circle of the cam. Some cams have so-called quietening ramps and this procedure affords the only reliable means of ensuring the tappet follower is positioned at minimum lift.

2 The exhaust valve clearance is 0.004 in set with the engine cold. A feeler gauge of this thickness should be a good sliding fit between the end of the valve stem and the rocker arm if the adjustment is correct. To reset the gap, slacken the locknut of the adjuster screw at the rocker arm tip and obtain the correct gap by varying the setting of the adjuster screw. When the gap is correct, tighten the locknut and recheck. Repeat this procedure for the right hand exhaust valve, rotating the engine so that the left hand exhaust valve is fully open. Replace both inspection caps, using new sealing washers.

3 The inlet valve clearance is 0.002 in set with the engine cold. Use an identical procedure to that adopted for the exhaust valve settings; when one valve clearance is being checked the corresponding valve in the other cylinder should be on full lift. Replace the inlet rocker box inspection caps.

4 Engines fitted with cast iron cylinder heads have reduced tappet clearances, 0.001 inch for all valves.

Fig. 1.9 This 'wheel' mark indicates the engine has ramp cams fitted. Mark is close to engine serial number

36.8 Tightening the dynamo retaining clamp screw

5   The 6T and late T110 models are fitted with ramp cams, which necessitate a 0.010 in valve clearance for both valves. Engines thus fitted have a special mark close to the engine serial number.

## 36 Engine reassembly - completion and final adjustments

1   Re-attach the rocker oil feed pipe, using new sealing washers on both side of the oil unions. Take care that the unions do not turn whilst the acorn nuts are tightened, otherwise there is risk of the small diameter pipe necking and cutting off the oil supply.
2   Refit the carburettor(s) taking care the heat insulator(s) is not omitted. It should be positioned between the flange of the carburettor(s) and the inlet manifold. Do not overtighten the nuts that retain the carburettor to the cylinder head or the flange will bow and give rise to a bad air leak. Refit the carburettor top(s), making sure the needle(s) engages correctly with the needle jet(s) and the air slide(s) with its guide cutaway in the jet block(s). Tighten the knurled ring(s) at the top of the carburettor(s). The thread is fine and they are easily cross-threaded.
3   Reconnect the ignition advance cable to the handlebar control, if the machine is fitted with a manually operated magneto. Check that all of the control cables have easy sweeps and are not likely to be trapped by the petrol tank, when it is refitted.
4   Reconnect the oil pipe junction block to the rear of the right-hand crankcase. Use a new gasket, without any gasket cement and make sure the gasket does not obstruct the oilways. Tighten the retaining bolt fully. Refill the oil tank with 5 pints of oil of the recommended viscosity, after checking the drain plug is back in position, and tight. Refill the primary chaincase with 1/3 pint SAE 20 oil (190 ccs) and replace the inspection cap after checking the primary chain tension.
5   Refit the complete exhaust system, making sure the exhaust pipes are clamped firmly on the exhaust stubs of the cylinder head. Do not omit the two stays that connect with the front engine plates.
6   Replace the petrol tank, making sure none of the control cables are trapped or bent at an awkward angle. Check that the controls operate smoothly. Reconnect the petrol pipes.
7   Remake the electrical connections, and reconnect the sparking plug leads. If the engine has been timed according to the recommendations, the lead nearest the engine should connect with the left-hand sparking plug and the opposite lead

to the one on the right. If a distributor is fitted, make sure the plastic protective sheet is fitted with the cutaway on the left of the engine. Fit the high tension leads from the distributor cap into this cutaway.
8   Replace the dynamo, using a new gasket at the joint with the back of the timing cover. Replace and tighten the retaining nut or screw, then tighten the screw in the retaining clamp around the body. Remake the electrical connections.

## 37 Starting and running the rebuilt engine

1   Start and run the engine slowly for the first few minutes, especially if the engine has been rebored. Remove the cap from the top of the oil tank and check that oil is returning. There may be some initial delay whilst the pressure builds up and oil circulates throughout the system, but if none appears after the first few minutes running, stop the engine and investigate the cause. The 'tell tale' oil indicator of the early models should extend from the timing cover if the oil pressure has built up satisfactorily. Stop the engine if it stays closed and investigate the cause.
2   Check that all controls function correctly and that the generator is indicating a charge on the ammeter. Check for any oil leaks or blowing gaskets.
3   Before taking the machine on the road, check that all the legal requirements are fulfilled and that items such as the horn, speedometer and lighting equipment are in full working order. Remember that if a number of new parts have been fitted, some running-in will be necessary. If the overhaul has included a rebore, the running-in period must be extended to at least 500 miles, making maximum use of the gearbox so that the engine runs on a light load. Speeds can be worked up gradually until full performance is obtainable by the time the running-in period is completed.
4   Do not tamper with the exhaust system under the mistaken belief that removal of the baffles or replacement with a quite different type of silencer will give a significant gain in performance. Although a changed exhaust note may give the illusion of greater speed, in a great many cases quite the reverse occurs in practice. It is therefore best to adhere to the manufacturer's specification.

## 38 Engine modifications and tuning

1   The Triumph twin engine can be tuned to give even higher performance and yet retain a good standard of mechanical reliability. Many special parts for boosting engine performance are available both from the manufacturer and from a number of specialists who have wide experience of the Triumph marque. The parts available include high compression pistons, high lift camshafts and even cylinder heads with four valves per combustion chamber.
2   There are several publications, including a pamphlet available from the manufacturer, that provide detailed information about the ways in which a Triumph twin engine can be modified to give increased power output. It should be emphasised, however, that a certain amount of mechanical skill and experience is necessary if an engine is to be developed in this manner and still retain a good standard of mechanical reliability. Often it is preferable to entrust this type of work to an acknowledged specialist and therefore obtain the benefit of his experience.
3   With certain reservations, it is not recommended to tune any pre-unit construction engine manufactured before 1955 since the smaller diameter built-up crankshaft assembly is less likely to withstand the extra stresses over a long period.

**39 Fault diagnosis: engine**

| Symptom | Cause | Remedy |
| --- | --- | --- |
| Engine will not turn over | Clutch slip | Check and adjust clutch. |
| | Mechanical damage | Check whether valves are operating correctly and dismantle if necessary. |
| Engine turns over but will not start | No spark at plugs | Remove plugs and check. Check magneto. Check whether battery is discharged. (alternator models only). |
| | No fuel reaching engine | Check fuel system. |
| | Too much fuel reaching engine | Check fuel system. Remove plugs and turn engine over several times before replacing. |
| Engine fires but runs unevenly | Ignition and/or fuel system fault | Check systems as though engine will not start. |
| | Incorrect valve clearances | Check and reset. |
| | Burnt or sticking valves | Check for loss of compression. |
| | Blowing cylinder head gasket | See above. |
| Lack of power | Incorrect ignition timing | Check accuracy of setting. |
| | Valve timing not correct | Check timing mark alignment on timing pinions. |
| | Badly worn cylinder barrel and pistons | Fit new **rings** and pistons after rebore. |
| High oil consumption | Oil leaks from engine gear unit | Trace source of leak and rectify. |
| | Worn cylinder bores | See above. |
| | Worn valve guides | Replace guides. |
| Excessive mechanical noise | Failure of lubrication system | Stop engine and do not run until fault located and rectified. |
| | Incorrect valve clearances | Check and re-adjust. |
| | Worn cylinder barrel (piston slap) | Rebore and fit oversize pistons. |
| | Worn big end bearings (knock) | Fit new bearing shells. |

# Chapter 2 Gearbox

## Contents

## Specifications

Overall gear ratios

| Model | 5T | T100 | TR5 | 6T | TR6 | T110 | T120 |
|---|---|---|---|---|---|---|---|
| Fourth gear | 5.0 | 5.0 | 5.24 | 4.57 | 4.57 | 4.57 | 4.57 |
| Third gear | 5.95 | 5.95 | 6.24 | 5.45 | 5.45 | 5.45 | 5.45 |
| Second gear | 8.45 | 8.45 | 8.85 | 7.75 | 7.75 | 7.75 | 7.75 |
| First gear | 12.20 | 12.20 | 12.80 | 11.20 | 11.20 | 11.20 | 11.20 |

Gearbox sprocket      --------------Standardised at 18 teeth --------------

## 1 General description

A four-speed constant mesh gearbox is fitted to all the 500cc and 650cc pre-unit construction Triumph twins, the terminology itself denoting the fact that the gearbox is a fitting separate from the engine. The design of the gearbox has changed in minor details only, since its inception during the late thirties.

Unlike most other British manufacturers, Triumph have remained faithful to the 'up for up' and 'down for down' method of gear selection, which can be changed to the more conventional mode of operation only if the gearchange pedal is reversed on its shaft so that the pedal faces rearwards.

Most operations can be accomplished with the gearbox in location and there is no necessity to remove it from the frame unless the shell itself is damaged or a complete overhaul is envisaged, where better access would prove advantageous. Removal of the gearbox necessitates a considerable amount of dismantling as a prelude, that includes removal of the exhaust system, footrests, brake pedal, primary chaincase, clutch and engine sprocket, clutch cable, speedometer cable, oil tank and the rear engine plates. In consequence, removal of the gearbox as a complete unit becomes a major operation, and should not be taken lightly.

## 2 Dismantling the gearbox - general

1 Before commencing work on the gearbox, make sure that the outer surfaces of the gearbox end covers and shell are clean and dry. Lay a sheet of clean paper immediately below the gearbox so that any parts inadvertently misplaced will fall onto a clean surface.

2 A good fitting screwdriver is essential to prevent damage to the screws retaining the outer end cover in position; also a slim socket or box spanner to fit the recessed nuts which complete the method of end cover retention. If the screw heads are damaged, it will prove difficult to remove or replace the screws, even when the correct screwdriver bit is available.

### 3  Dismantling the gearbox - removing the outer end cover

1  Before access can be gained to the outer end cover, it is necessary to remove the right hand exhaust pipe, silencer and the right hand footrest. Slacken the finned clip fitted to the right hand exhaust pipe and remove the bolts from the exhaust pipe stay and from the lug of the silencer. If the exhaust system is 'siamesed', slacken the clamp around the joint between both systems. The exhaust pipe can now be freed by a few taps with a rawhide mallet and the complete right hand system removed. The footrest is secured by a nut and washer on the end of the footrest rod.

2  Slacken off the clutch cable adjuster at the gearbox end and detach the outer end of the cable from the external operating arm on the end of the gearbox outer cover. The adjuster can now be unscrewed from the top of the gearbox inner cover and the clutch cable removed completely from the vicinity of the gearbox. Detach the speedometer drive cable by unscrewing the union nut at the base of the gearbox inner cover.

3  Place a tray beneath the gearbox and unscrew the gearbox drain plug. The gearbox holds just under one pint of oil.

4  Engage top gear. Although not strictly necessary, it will facilitate the removal of several nuts which have to be removed during the dismantling sequences to follow by permitting the engine to be locked, when the rear brake is applied.

5  Unscrew the top and bottom hexagonal nuts and the five screws from the end cover. Depress the kickstarter slightly, to clear it from the top, and tap the end cover until it can be drawn away from the gearbox shell.

6  There is no necessity to remove either the gear change lever or the kickstarter unless attention is required to either mechanisms.

### 4  Dismantling the gearbox - removing the inner end cover and gear clusters

1  Removal of the inner cover may be hampered by the flexible oil pipes immediately below the gearbox. If it is necessary to remove them, the oil tank must first be drained. It is recommended that the pipes are detached by removing the union at the point where the pipes join the crankcase, which is more accessible after the gearbox outer cover has been removed. The union is retained to a stud by one centre nut and washer; there is

4.2 Detach kickstart ratchet mechanism to release inner cover

4.2a The inside of the inner cover, in an inverted position

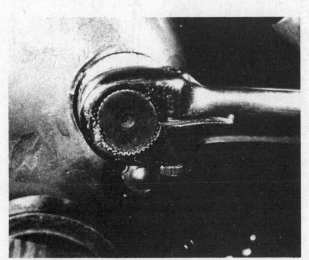

3.6 No necessity to detach gearchange lever

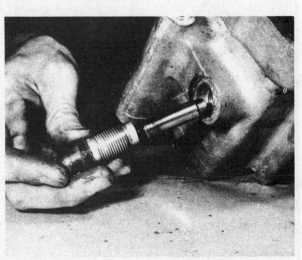

4.4 Unscrew domed nut to release camplate plunger

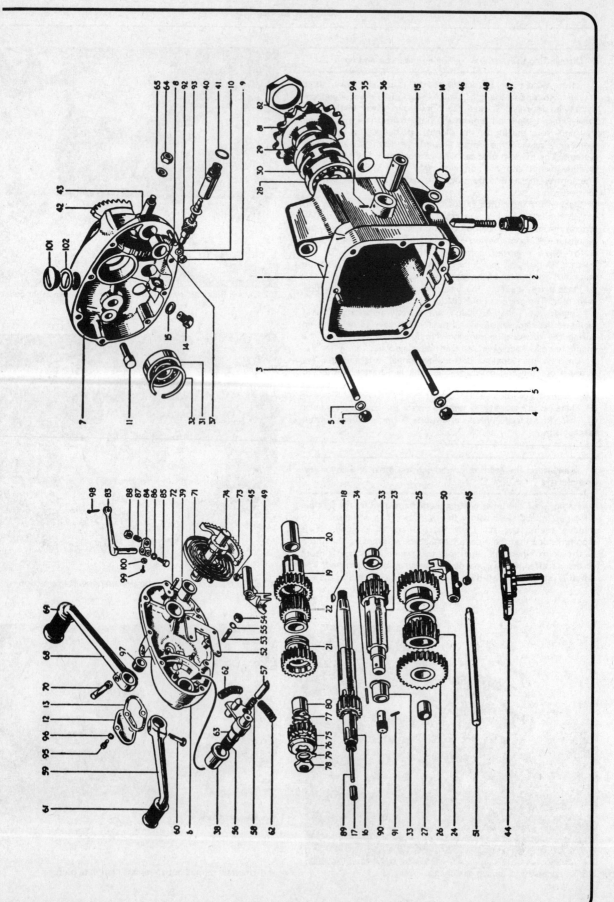

**Fig. 2.1 Triumph 4-speed gearbox**

1 Gearbox shell
2 Dowel - 2 off
3 Stud - 2 off
4 Nut - 2 off
5 Washer - 2 off
6 Outer cover
7 Inner cover
8 Stud
9 Nut
10 Washer
11 Kickstart stop
12 Inspection cover
13 Gasket
14 Level and drain plug
15 Washer
16 Mainshaft
17 Bush
18 Key
19 Top gear pinion, mainshaft
20 Bush
21 Third gear pinion, mainshaft
22 Second gear pinion, mainshaft
23 Layshaft
24 Third gear pinion, layshaft
25 Second gear pinion, layshaft

26 Bottom gear pinion, layshaft
27 Bush
28 Mainshaft sleeve gear bearing
29 Oil seal
30 Circlip
31 Mainshaft bearing (right-hand)
32 Circlip
33 Layshaft bush
34 Peg
35 Blanking disc
36 Camplate bush
37 Gearchange spindle bush
38 Gearchange spindle bush
39 Kickstart bush
40 Speedometer drive bush
41 Sealing ring
42 Selector quadrant
43 Spindle
44 Camplate
45 Roller - 2 off
46 Indexing plunger
47 Domed nut
48 Spring
49 Mainshaft selector fork
50 Layshaft selector fork

51 Selector fork rod - 2 off
52 Plunger guide plate
53 Stud - 4 off
54 Nut - 4 off
55 Washer - 4 off
56 Gearchange quadrant
57 Gearchange plunger
58 Spring
59 Gearchange pedal
60 Bolt
61 Gearchange pedal rubber
62 Pedal return spring - 2 off
63 O ring
64 Nut
65 Washer
68 Kickstart crank
69 Kickstart rubber
70 Kickstart cotter pin
71 Kickstart return spring
72 Spring anchor peg
73 Kickstart quadrant
74 Kickstart spindle
75 Pinion
76 Ratchet
77 Ratchet spring

78 Nut
79 Tab washer
80 Sleeve
81 Final drive sprocket
82 Nut
83 Clutch operating lever
84 Clutch operating arm
85 Peg
86 Serrated washer
87 Clutch rod adjuster
88 Lock nut
89 Speedometer drive pinion
90 Speedometer drive pinion
91 Peg
92 Speedometer drive gear
93 Thrust washer
94 Camplate bush peg
95 Screw
96 Washer
97 Kickstart spindle rubber
98 Split pin
99 Peg
100 Clutch operating roller
101 Filler plug
102 Washer

4.5 The selector spindle and forks

4.6 Unscrew nut on end of sleeve gear pinion and ....

4.6a Pull sprocket off splines

5.1 Sleeve gear bearing is retained by a circlip

5.4 Layshaft bearings are pegged

8.3 Renew kickstart return spring if weak or damaged

8.4 Renew ratchet pinion at same time as ratchet quadrant

9.1 Gearchange mechanism rarely requires attention

10.1 Check clutch operating mechanism for wear

no chance of the pipes being inadvertently reversed if this procedure is adopted.

2   Before the inner cover can be removed, it will be necessary to remove the kickstarter pinion ratchet retaining nut from the end of the gearbox mainshaft, after bending back the tab washer. It can be locked by applying the back brake whilst the gearbox is held in top gear. Draw the ratchet assembly off the mainshaft complete with spring and inner bush. Withdraw the clutch pushrod from the centre of the mainshaft. Remove the single screw in the inside of the inner cover and the two nuts and washers at the rear. The inner cover will now pull off as the end of the mainshaft is lightly tapped through the end of the main bearing.

3   If it is necessary to remove the final drive sprocket, to renew the gearbox main bearing or oil seal, it is advisable to remove the outer chaincase cover at this stage and dismantle the primary drive, as described in Chapter 1, Sections 5.13 - 5.25. It is much more difficult to remove these components at a later stage since there will be no support for the gearbox mainshaft and layshaft after the inner cover is removed.

4   Unscrew the large domed nut from beneath the gearbox and withdraw it complete with the camplate plunger and spring. Then remove the gearbox inner cover. A few light taps with a rawhide mallet may be necessary to displace the cover so that it can be slid off, over the end of the retaining studs.

5   Remove the selector fork spindle, which is a push fit into the gearbox shell. The gearbox mainshaft can now be withdrawn together with the lower gear pinions and the selectors. The layshaft and the remaining gear pinions can be withdrawn, leaving only the camplate assembly and the mainshaft top gear. The camplate will pull off its shaft without difficulty. Do not lose the rollers from the selector forks.

6   To remove the mainshaft top gear (sleeve gear pinion) it is necessary to bend back the tab washer before unscrewing the large retaining nut (right hand thread). The nut measures 1.66 in across the flats and can be removed by Triumph service tool Z63 if a spanner of a suitable size is not available. When the nut is removed, the sleeve gear pinion is released by driving it into the gearbox shell with a hammer and soft metal drift. This action will free the final drive sprocket from the splines.

## 5   Dismantling the gearbox - removing (and replacing) the mainshaft and layshaft bearings

1   The left hand sleeve gear bearing is retained in the gearbox shell by a circlip which is preceded by an oil seal. Since it is necessary to renew the oil seal whenever the bearing is disturbed, prise the old seal out of position and remove the circlip with a pair of long nosed pliers.

2   Heat the gearbox shell locally with a blow lamp and drive the bearing out from the inside of the gearbox using a suitable drift. Before the shell cools, fit the replacement bearing and make sure it is pressed to the full depth of the housing before the circlip is fitted. Fit the new oil seal, after the shell has cooled.

3   The right hand mainshaft bearing is also retained by a circlip in the gearbox inner end cover. Heat treatment will again be required to help free the bearing after the circlip has been removed. The replacement bearing should be located before the end cover is allowed to cool and the circlip replaced.

4   The layshaft is supported by a plain bush at each end. If the castings are warmed, the bushes can be driven out of position and new replacements fitted whilst the castings are still warm. There is no necessity to remove the pegs, although if they happen to come out with the bushes, they must be replaced. Do not omit the blanking plate behind the left-hand bush. It must provide a perfect oil seal.

## 6   Slickshift and late-type gearboxes

1   Certain differences will be found in the Slickshift and later

type gearboxes, as the result of the redesign of the outer cover to give a cleaner and more stylish appearance. In the main, these relate to the repositioning of the external clutch operating arm, a filler plug in the top of the gearbox inner cover, and in the case of the Slickshift, the linking of the gearchange pedal to the clutch operating mechanism so that the two operations are interconnected. Internally, the main gearbox components are identical.

2   Attention to the Slickshift mechanism is described in a later Section of this Chapter.

## 7   Examination and renovation - general

1   Before the gearbox is reassembled, it will be necessary to inspect each of the components for signs of wear or damage. Each part should be washed in a petrol/paraffin mix to remove all traces of oil and metallic particles which may have accumulated as the result of general wear and tear within the gearbox.

2   Do not omit to check the castings for cracks or other signs of damage. Small cracks can often be repaired by welding, but this form of reclamation requires specialist attention. Where more extensive damage has occurred it will probably be cheaper to purchase a new component or to obtain a serviceable secondhand part from a breaker.

3   If there is any doubt about the condition of a part examined, especially a bearing, it is wise to play safe and renew. A considerable amount of strip down work will be required again if the part concerned fails at a much earlier date than anticipated.

## 8   Kickstarter mechanism - examination and renovation

1   If the kickstarter quadrant shows signs of wear, it should be renewed, otherwise it will tend to jam during initial engagement with the ratchet pinion. Note that the first tooth of the quadrant is relieved to minimise the risk of a jam when the initial engagement is made.

2   The quadrant is a drive fit on the splines of the kickstarter shaft. The kickstarter is removed by unscrewing the nut on the end of the cotter pin and driving out the pin with a hammer. When replacing a quadrant, it is important to ensure that the flat in the shaft is positioned correctly in relation to the quadrant or the operating angle will be incorrect. A new oil seal should be fitted over the shaft in the outer face of the end cover.

3   Renew the return spring if the action is weak or if the spring has suffered damage. The accompanying diagram shows the correct location of the spring prior to tensioning.

4   If the kickstarter ratchet has to be renewed, it is probable that the ratchet pinion, with which it engages, will require renewal too. It is bad policy to run old and new parts together.

5   It is also important to check the condition of the ratchet teeth and to renew both parts of the ratchet system if the teeth are rounded at their edges. A worn ratchet will eventually slip and make starting difficult. Renew the light return spring if it has taken a permanent set or has reduced return action.

## 9   Gearchange mechanism - examination and renovation

1   It should not be necessary to dismantle the gear change mechanism unless the gear lever return springs have broken or wear of the operating mechanism is suspected on account of imprecise gear changes. To dismantle the assembly, detach the gear change lever by slackening the pinch bolt and pulling the lever off the splined shaft. Remove the four nuts and

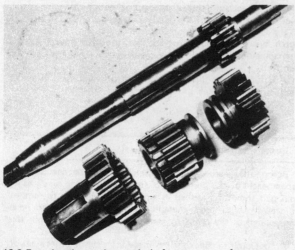

12.2 Examine the gearbox mainshaft components for wear or damage ....

12.2a .... followed by the layshaft components

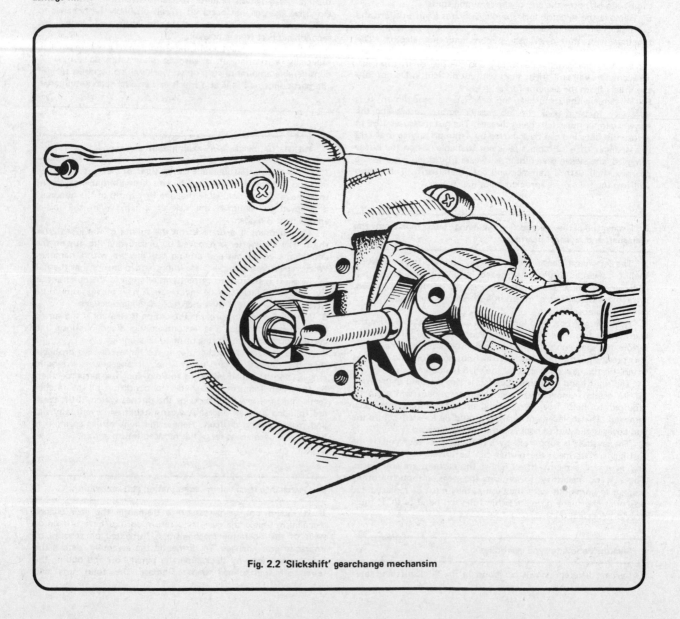

Fig. 2.2 'Slickshift' gearchange mechansim

lockwashers securing the guide plate to the inside of the outer end cover and withdraw the guide plate complete with plunger quadrant and curved gear change lever return springs.

2  Examine the various components for wear, especially the gear change plungers and the plunger springs. Each spring should have a free length of 1¼ in; renew them if they have taken a set. The plungers must be a clearance fit in the quadrant if they are to function correctly.

3  If the plunger guide plate is worn or grooved on the taper guide surfaces it must be renewed.

4  The gear change lever return springs seldom give trouble unless they become fatigued or if condensation within the gearbox causes corrosion. If there is any doubt about their condition renew them as a matter of course.

5  After a lengthy period of service the gear change quadrant bush may wear oval. If this form of wear is evident, the bush must be renewed.

6  If the teeth of the camplate operating quadrant attached to the inner end cover are chipped, indented or worn, the quadrant must be renewed. It is retained by two split pins which, when removed, will permit the spindle to be withdrawn.

7  Do not omit to check the oil seal around the shaft of the gear change lever in the outer end cover.

12.6 Remove camplate to check for wear

## 10 Clutch operating mechanism - examination and renovation

1  The clutch operating mechanism, attached to the outer end cover by two screws, seldom gives trouble provided the gearbox oil level is maintained.

## 11 The Slickshift gearchange mechanism

1  As the accompanying illustration will show, the way in which the gearchange pedal and clutch are interconnected is both simple and ingenious. There is no means of adjustment; most cases of faulty gear selection can be attributed to one of three causes, a badly adjusted clutch, a sticking camplate plunger or slackness of the nuts on the gearbox mainshaft.

2  If the ramps of the gearchange quadrant are worn and/or the extension of the clutch operating arm that runs on these ramps, the parts concerned will have to be renewed. This type of wear is more likely to occur if the gearbox is run for long periods with a low oil content.

3  On Slickshift and late type gearboxes, it is necessary to remove the oval inspection plate of the gearbox end cover in order to detach the end of the clutch cable from the operating arm. The plate is held by two screws and has a sealing gasket.

12.6a This camplate has worn badly and had to be renewed

## 12 Gearbox components - examination and renovation

1  Examine each of the gear pinions carefully for chipped or broken teeth. Check the internal splines and bushes. Instances have occurred where the bushes have worked loose or where the splines have commenced to bind on their shafts. The two main causes of gearbox troubles are running with low oil, and condensation, which gives rise to corrosion. The latter is immediately evident when the gearbox is dismantled.

2  The mainshaft and the layshaft should both be examined for fatigue cracks, worn splines or damaged threads. If either of the shafts have shown a tendency to seize, discolouration of the areas involved should be evident. Under these circumstances check the shafts for straightness.

3  Harsh transmission is often caused by rough running ball races especially the mainshaft ball journal bearings. Section 5 of this Chapter describes the procedure for removing and replacing the gearbox bearings.

4  All gearbox bearings should be a tight fit in their housings. If a bearing has worked loose and has revolved in its housing, a

14.1 Fit new oil seal to sleeve gear pinion

bearing sealant such as Loctite can be used, provided the amount of wear is not too great.

5 Check that the selector forks have not worn on the faces which engage with the gear pinions and that the selector fork rod is a good fit in the gearbox housings. Heavy wear of the selector forks is most likely to occur if replacement of the mainshaft bearings is long overdue.

6 The gear selector camplate will wear rapidly if the mainshaft bearings need replacement. Although the gear pinion behind the camplate is unlikely to wear excessively, it should be inspected if it has proved difficult to select gears.

7 The camplate plunger must work freely within its housing. Check the free length of the spring, which should be 2½ in if it has not compressed.

8 Play, accompanied by oil leakage, is liable to occur if the bush within the sleeve gear pinion is worn. The working clearance is normally from 0.003 in to 0.005 in.

### 13 Gearbox reassembly - general

1 Before commencing reassembly, check that the various jointing surfaces are clean and undamaged and that no traces of old gasket cement remain. This check is particularly important because the gearbox will not remain oiltight if these simple precautions are ignored.

2 Check that all threads are in good condition and that the locating dowels, where fitted, are positioned correctly. Have available an oil can filled with clean engine oil so that the various components can be lubricated during reassembly.

### 14 Gearbox reassembly - replacing the sleeve gear pinion and final drive sprocket

1 Assuming the mainshaft main bearing is in position in the extreme left hand end of the gearbox shell, and retained by a circlip, drive the new oil seal into position with the lip and spring innermost. Lubricate the sleeve gear pinion and drive it through the main bearing from inside the gearbox, taking special care that the feather edge of the new oil seal is not damaged.

2 Lubricate the ground, tapered boss of the final drive sprocket with oil and slide it on to the splines of the sleeve gear. Replace the tab washer and then the large retaining nut which should be fingertight at this stage. Reconnect the final drive chain (closed end of the spring link facing the direction of the travel of the chain) and apply the rear brake so that the sprocket retaining nut can be tightened fully. Bend over the tab washer so that the nut is retained in place.

### 15 Gearbox reassembly - replacing the camplate, camplate plunger, gear pinions and selectors

1 Lubricate the camplate spindle and replace the camplate in its correct location within the gearbox. Assemble the camplate plunger and spring into the domed retaining nut, and check the plunger moves quite freely when it is within its housing. Place a new fibre washer above the domed nut and screw the plunger assembly into the bottom of the gearbox. Tighten the nut and check that the camplate is positioned so that the plunger engages with the 4th (top) gear notch. This position must be maintained whilst the gearbox is reassembled, otherwise the gears will not index correctly.

2 Lubricate the gear pinions attached to the mainshaft and layshaft, then assemble the mainshaft and layshaft gear clusters. Grease the camplate rollers to retain them on the selector forks and position the forks. Note that the fork with the smaller radius must engage with the mainshaft gear cluster.

3 Insert the gear train and selectors as a complete unit and ensure that the camplate rollers engage correctly with the roller track in the camplate. This is a somewhat tricky operation and it

may be necessary to make several attempts before all of the components locate correctly.

4 When assembly is complete, lubricate the selector fork spindle and insert it through the selector forks after checking that the bores are approximately in line. The shouldered end of the rod should be inserted first and the shaft pushed home until it engages fully with the gearbox housing. Check that the mainshaft selector forks is in the innermost position.

### 16 Gearbox reassembly - replacing the inner end cover and indexing the gears

1 Use the oil can to lubricate all parts at present located within the gearbox shell. It is advisable to replace the gearbox drain and level plug first to prevent subsequent leakage.

2 Apply a thin coating of gasket cement to the face joint of the gearbox shell and to the back of the inner end cover which is to mate with it. Replace the two dowels at the top and bottom of the gearbox shell face joint and check that the camplate is still positioned so that the plunger is in full engagement with the fourth (top) gear notches. Slide the inner end cover over the two projecting studs of the gearbox shell and, when it is only about ¼ inch from the jointing face, position the camplate quadrant so that it is in the fourth gear position. Push the cover fully home and again check that the alignment of the camplate quadrant is correct; the centre tooth of the quadrant should be in line with the gearbox mainshaft if correctly positioned. The need for careful alignment cannot be overstressed since this determines the correct indexing of the gears. If the alignment is as much as a tooth out, the gear change lever will lock in position during gear changing, necessitating a further strip down to rectify the alignment.

3 Replace and tighten the screws and two bolts which retain the inner end cover to the gearbox shell.

4 Replace the oil pipe union over the stud near the base of the engine timing cover, using a new gasket, and tighten the nut (and washer) holding the union in place. Refill the oil tank.

5 Temporarily replace the outer end cover and check that the gear change sequence is correct by operating the gear change lever and turning the rear wheel in unison. Any gear selection problems must be remedied at this stage by adjusting the position of the camplate quadrant in relation to that of the pre-set camplate.

6 When the gear change sequence is correct, remove the outer end cover and replace the kickstarter pinion and ratchet assembly, not forgetting the thin centre bush and the return spring. Place the tab washer and the securing nut on the end of the mainshaft and tighten the nut securely whilst applying the

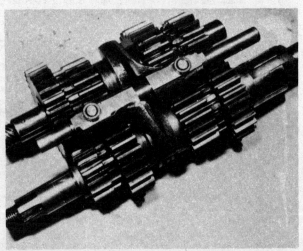

15.2 Make trial assembly of gear cluster and selectors on bench

15.3 Insert gear cluster with selectors, minus selector fork spindle

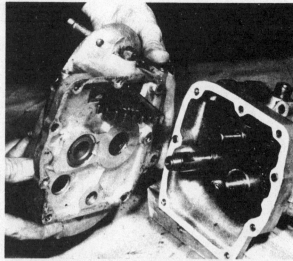

16.2 Fit new gasket when replacing the inner cover

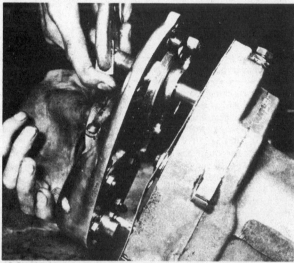

16.6 If kickstart is detached, return spring must be pre-tensioned

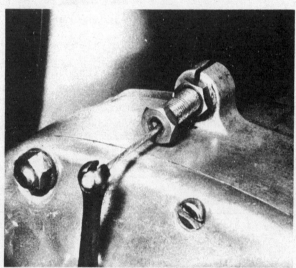

17.3 Make fine adjustments here to ensure slack in clutch cable

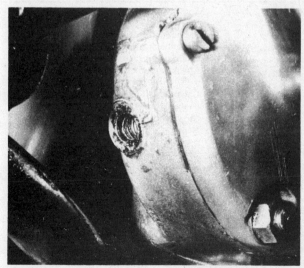

17.4 Refill gearbox until oil emerges from level hole

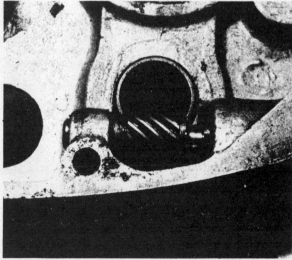

18.1 This pinion needs changing if gear ratios are altered

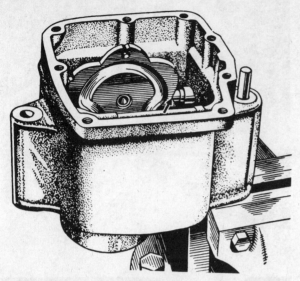

**Fig. 2.3 Correct positioning of gearchange camplate on reassembly**

*Camplate plunger position shown by arrow*

back brake to prevent the mainshaft from turning. Bend the tab washer to lock the nut in position. Note that overtightening the nut may result in distortion or failure of the thin centre bush within the kickstarter pinion assembly. Re-insert the clutch pushrod.

7  Reconnect the final drive chain (closed end of the spring link facing the direction of the travel of the chain) and apply the rear brake so that the sprocket retaining nut can be tightened fully. Bend over the tab washer so that the nut is retained in place.

## 17 Gearbox reassembly - replacing the outer end cover and completing assembly

1  Coat the jointing surfaces of the outer and inner end covers with a thin layer of gasket cement and refit the end cover, after checking that the two locating dowels are in place. The kickstarter return spring must be tensioned one turn before the cover is placed in position, by rotating the kickstarter in an anticlockwise direction and holding it under tension until the cover is fully home. Note that it will be necessary to depress the kickstarter a further half turn whilst the cover is positioned so that the quadrant will clear the end stop. Replace the crosshead screws and the nuts to retain the cover in position.

2  Check that the kickstarter returns correctly and that all of the gears are selected in the correct sequence. It will be necessary to turn the rear wheel when making this latter check to ensure the gear pinions engage to their full depth.

3  Replace the primary drive by following the procedure given in Chapter 1, Section 34 and adjust the clutch springs before the chaincase outer cover is replaced. Reconnect the clutch cable to the gearbox, engage the nipple with the clutch operating arm and adjust for a small amount of slack in the cable.

4  Refill the gearbox with 2/3 pint (380 cc) SAE 50 oil and the primary chaincase with 1/3 pint (190 cc) SAE 20 oil.

5  Replace the gear change lever (if removed) and reconnect the speedometer drive cable. Replace the right hand footrest and the right hand exhaust system.

## 18 Changing speedometer drive gear combinations

1  Early machines have the speedometer drive taken from the right hand end of the gearbox layshaft and if any change is made in the overall gear ratios by changing the size of the gearbox final drive sprocket, as for sidecar work, the speedometer drive ratio has to be corrected in order to preserve accuracy.

2  It will be noted that only the final drive sprocket has to be changed when a sidecar is attached to the machine in order to achieve the optimum gear ratios.

3  The layshaft drive pinion is a push fit in the right hand end of the layshaft hollow and is secured with a pin that passes through the layshaft.

## 19 Fault diagnosis - gearbox

| Symptom | Cause | Remedy |
| --- | --- | --- |
| Difficulty in engaging gears | Gears not indexed correctly | Check timing sequence of inner end cover (will occur only after rebuild). |
| | Worn or bent gear selector forks | Examine and renew if necessary. |
| | Worn camplate | Examine and renew as necessary. |
| | Low oil content | Check gearbox oil level and replenish. |
| Machine jumps out of gear | Mechanism not selecting positively | Check for sticking camplate plunger or gear change plungers. |
| | Sliding gear pinions binding on shafts | Strip gearbox and ease any high spots. |
| | Worn or badly rounded internal teeth in pinions | Replace all defective pinions. |
| Kickstarter does not return when engine is started or turned over | Broken kickstarter return spring | Remove outer end cover and replace spring. |
| | Kickstarter ratchet jamming | Remove end cover and renew all damaged parts. |
| Kickstarter slips on full engine load | Worn kickstarter ratchet | Remove end cover and renew all damaged parts. |
| Gear change lever fails to return to normal position | Broken or compressed return springs | Remove end cover and renew return springs. |

# Chapter 3 Clutch

**Contents**

**Specifications**

## Clutch

| | |
|---|---|
| Number of inserted plates ... ... ... ... ... ... | 5 |
| Number of plain plates ... ... ... ... ... ... | 6 |
| Pressure springs ... ... ... ... ... ... ... | 4 |
| Free length ... ... ... ... ... ... ... ... | 1.5 inches (3.81 cm) |
| Clutch rollers - diameter ... ... ... ... ... ... | 0.2500 - 0.2495 inches (6.350 - 6.337 mm) |
| Clutch chainwheel: | |
|     number of teeth ... ... ... ... ... ... ... | Standardised at 43 |
| Primary chain ... ... ... ... ... ... ... | ½ inch x 0.305 inches, 79 rollers - all models |
| Primary chaincase: | |
|     oil content ... ... ... ... ... ... ... | 1/3 pint SAE 20 oil (190 cc) |

## 1 General description

1 The clutch is of the multi-plate type, designed to operate in oil. Cork is used to line the inserted plates of the early models; a change was made to a synthetic friction material at a later stage.
2 Models having an alternator on the end of the crankshaft have a shock absorber within the centre drum of the clutch, to absorb and even out transmission surges. Earlier models have a shock absorber of the spring and cam type incorporated in the engine sprocket assembly.
3 Drive from the engine to the separate gearbox is by means of a chain which may or may not be of the endless type. It is tensioned by tilting the gearbox about a cast-in lug in the base of the gearbox shell, so that adjustments can be made at regular intervals to take up wear.

## 2 Adjusting the clutch

1 The clutch can be adjusted by means of the cable adjuster, in the top of the gearbox end cover, the pushrod adjuster in the centre of the clutch pressure plate and by varying the tension of the four clutch springs. In the latter case, the chaincase cover must be removed before adjustment can be effected.
2 To adjust the clutch, slacken off the gearbox adjuster so that there is an excess of free play in the lever, then unscrew the locknut of the adjuster at the end of the clutch operating arm and turn the screw so that there is approximately 1/16 inch free play. Tighten the locknut and recheck the clearance, then readjust the cable adjuster in the top of the gearbox until the free play in the cable is increased to 1/8 inch.

3 If the adjustment is correct and none of the components is worn, the external operating lever should lie approximately parallel to the gearbox end cover as the clutch is withdrawn (later models) or be near horizontal (early models). If the angle is not correct, clutch action may be heavy since the maximum

2.2 Main adjustment is at clutch operating arm

2.6 Check slots in outer drum for indentations, also ....

2.6a .... slots in clutch centre

3.2 Examine clutch plates carefully to determine extent of wear

3.4 Clutch springs will compress with use. Check length

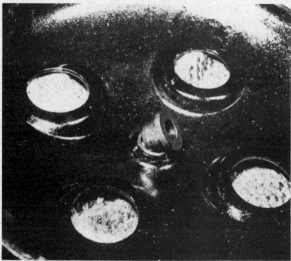

3.5 Check hardened insert for wear

3.5a .... also ends of pushrod

leverage is not available. The usual cause is a worn clutch pushrod.

4 If the clutch still drags and the adjustment procedure described has produced no improvement, it will be necessary to remove the primary chaincase cover, in order to gain access to the four clutch spring adjusters in the pressure plate. Before slackening the adjuster nuts, check that the drag is not caused by uneven tensioning of the pressure plate. This check is made by using the handlebar lever to slip the clutch and turning the clutch by depressing the kickstarter. Uneven tension will immediately be obvious by the characteristic 'wobble' of the pressure plate. The wobble can be corrected by retensioning the clutch springs individually until the pressure is even.

5 If the clutch still drags, the adjusters should be slackened off an equal amount at the time before a recheck is made. Do not slacken them off too much, or the clutch operation will become very light with the possibility of clutch slip when the engine is running.

6 Drag is often caused by wear of the clutch outer drum. The projecting tongues of the inserted plates will wear notches in the grooves of the drum, which will eventually trap the inserted plates as the clutch is withdrawn and prevent them from separating fully. In a case such as this, renewal of both the clutch chainwheel and the inserted clutch plates is necessary. If wear is detected in the early stages, it is possible to redress the grooves with a file until they are square once again, and to remove the burrs from the edges of the clutch plate tongues.

7 Heavy clutch operation is sometimes attributable to a cable badly in need of lubrication, or one in which the outer covering has become badly compressed through being trapped. Even a sharp bend in the cable will stiffen up the operation.

8 Clutch slip will occur when the clutch linings reach their limit of wear. If the reduction in linings thickness exceeds 0.030 in slip will occur and the inserted plates must be renewed. This necessitates dismantling the clutch, as described in Chapter 1, section 5.17. There is no necessity to dismantle any part of the generator or primary transmission assembly because the clutch plates can be withdrawn after the clutch pressure plate is removed.

9 It cannot be overstressed that a great many clutch problems are caused by failure to maintain the oil content of the chaincase at the correct level or by the use of a heavier grade of oil than that recommended by the manufacturers. It is also important that the chaincase oil is changed at regular intervals, to offset the effects of condensation.

## 3 Examining the clutch plates and springs

1 When the clutch is dismantled for replacement of the clutch plates or during the course of a complete overhaul, this ia an opportune time to examine all clutch components for signs of wear or damage.

2 As mentioned previously, the inserted clutch plates will have to be renewed when the reduction in thickness of the linings reaches 0.030 in. It is important to check the condition of the tongues at the edge of each plate which engage with the grooves in the clutch chainwheel. Even if the clutch linings have not approached the wear limit, it is advisable to renew the inserted plates if the width of the tongues is reduced as a result of wear.

3 Both plain and inserted plates should be perfectly flat. Check by laying them on a sheet of plate glass. Discoloration will occur if the plates have overheated; the surface finish of the plates is not too important provided it is smooth and the plates are not buckled.

4 The clutch springs will compress during service and take a permanent set. If any spring has reduced in length by more than 0.1 in, all four springs should be renewed - never renew one spring on its own. See Specifications section for the original free length measurement.

5 Check the hardened end of the clutch adjuster in the centre of the clutch pressure plate to ensure it has not softened from overheating or that the hardened surface is not chipped, cracked or worn away. This also applies to the ends of the clutch pushrod. Mysterious shortening of the pushrod, necessitating frequent clutch adjustment, can usually be attributed to incorrect adjustment of the clutch, such that the absence of free play in the clutch cable places a permanent load on the pushrod. This in turn causes the ends of the pushrod to overheat and soften, thus greatly accelerating the rate of wear.

6 The teeth of the clutch chainwheel should be examined since chipped, hooked or broken teeth will lead to very rapid chain wear. It is not possible to reclaim a worn chainwheel; the whole assembly must be renewed.

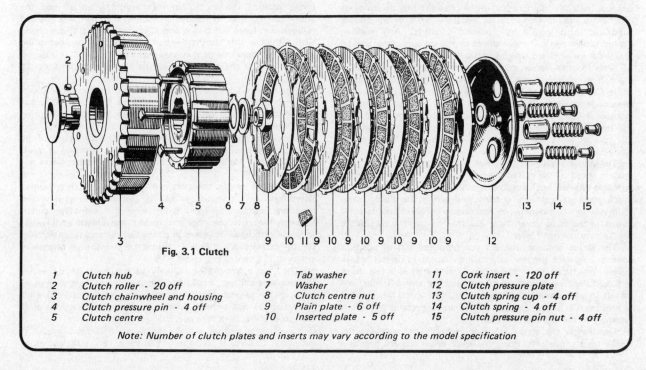

**Fig. 3.1 Clutch**

| | | | | | | |
|---|---|---|---|---|---|---|
| 1 | Clutch hub | 6 | Tab washer | 11 | Cork insert - 120 off |
| 2 | Clutch roller - 20 off | 7 | Washer | 12 | Clutch pressure plate |
| 3 | Clutch chainwheel and housing | 8 | Clutch centre nut | 13 | Clutch spring cup - 4 off |
| 4 | Clutch pressure pin - 4 off | 9 | Plain plate - 6 off | 14 | Clutch spring - 4 off |
| 5 | Clutch centre | 10 | Inserted plate - 5 off | 15 | Clutch pressure pin nut - 4 off |

*Note: Number of clutch plates and inserts may vary according to the model specification*

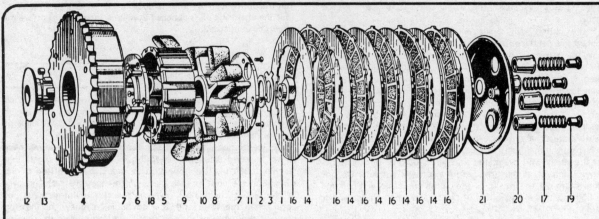

**Fig. 3.2 Clutch (shock absorber type)**

| | | | | | |
|---|---|---|---|---|---|
| 1 | Clutch centre nut | 8 | Shock absorber spider | 16 | Plain plate - 6 off |
| 2 | Washer | 9 | Driving insert - 4 off | 17 | Clutch spring - 4 off |
| 3 | Tab washer | 10 | Rebound insert - 4 off | 18 | Clutch pressure pin - 4 off |
| 4 | Clutch chainwheel and housing | 11 | Outer shock absorber retaining plate | 19 | Clutch pressure pin nut - 4 off |
| 5 | Clutch centre | 12 | Clutch hub | 20 | Spring cup - 4 off |
| 6 | Inner shock absorber retaining plate | 13 | Clutch roller - 20 off | 21 | Clutch pressure plate |
| 7 | Screw - 4 off | 14 | Inserted plate - 5 off | | |

7   The clutch operating mechanism at the right hand side of the gearbox is unlikely to give trouble, or wear rapidly.

---

#### 4   Examining the shock absorber assembly

1   Unless the machine is fitted with an alternator, the shock absorber assembly will be found on an extension of the left-hand crankshaft, in front of the engine sprocket. The engine sprocket has a boss projecting outwards with some cam-like contours on its face, the sprocket being free to revolve on a plain portion of the crankshaft. A collar, with matching contours, is fitted over the splined portion of the crankshaft and is kept in contact with the engine sprocket by means of a large compression spring. Because the collar is connected to the crankshaft by means of its internal splines, drive will take place all the time the engine sprocket and collar are pressed together. Any sudden transmission surge will make the cams rise up each other and so give the drive a certain amount of flexibility.

2   After lengthy service, the compression spring may weaken or the splines of either the collar and/or the crankshaft become worn. Compare the spring with a new one; if it has compressed, renewal is advisable. Never pack out the spring with washers to increase the tension, since this may cause the spring to become coil bound before the collar and engine sprocket can separate. In the case of worn splines, the parts concerned will have to be renewed.

3   The above arrangement is not possible when an alternator rotor is fitted to the end of the crankshaft and in consequence a different type of shock absorber is built in the clutch centre. The shock absorber assembly is contained within the clutch inner drum and can be examined by unscrewing the three countersunk screws in the front cover plate. Remove the cover plate using a small screwdriver as a lever.

4   The shock absorber rubbers can be prised out of position commencing with the smaller rebound rubbers, to make the task easier. Avoid damage to the rubbers which may disintegrate in service if they are punctured or cracked. The centre 'spider' will be left in position and need not be disturbed unless it is cracked or broken. It is held in position by the nut that retains the clutch assembly to the gearbox mainshaft.

5   Reassemble the shock absorber assembly by reversing the dismantling procedure. Insert the large drive rubbers first and

follow up with the smaller rebound rubbers. Fitting will be made easier if the smaller rubbers are smeared with household liquid detergent so that they can be slid into position. It is advisable to keep the rubbers free from oil, even though they are made of a synthetic material.

6   Before the cover plate screws are replaced, smear their threads with a sealant such as Loctite and tighten them fully. It is permissible to caulk the heads in position with a centre punch.

---

#### 5   Engine sprocket - removal and replacement

1   As mentioned in the previous Section, the engine sprocket forms part of the shock absorber assembly on all but the alternator models. To remove the engine sprocket, slacken and remove the sleeve nut on the end of the crankshaft; it is recessed within the cap that retains the end of the shock absorber spring and a slim socket or box spanner will be required. It may be necessary to jar the nut against engine compression, in order to start it.

2   When the nut has been unscrewed, lift off the cup, the shock absorber spring and splined collar. Before the engine sprocket can be withdrawn, it will first be necessary to detach the spring link from the chain and separate the chain.

3   When replacing the sprocket, make sure the spring link of the chain is secure, with the closed end facing the direction of travel of the chain.   Make sure also that the shock absorber sleeve nut is tightened fully.

4   Before the engine sprocket can be removed from an engine fitted with an alternator, it will be necessary to remove the generator stator coils and rotor, and the complete clutch assembly because the engine sprocket and clutch chainwheel must be withdrawn together as a result of the endless chain used for the primary drive. Full details of the dismantling procedure are given in Chapter 1, section 5.15.

5   The engine sprocket is unlikely to require attention unless the teeth are chipped, hooked or broken - unlikely unless a chain breakage has caused the engine to lock up. Reference to Chapter 1, section 34 will show how to replace the primary drive after the sprocket has been renewed.

6   Check the tension of the primary chain and adjust if necessary. It is easier to do so at this stage, before the chaincase

outer cover is fitted. Do not omit to refill the chaincase with oil: 1/3 pint (190ccs) SAE 20 oil.

## 6 Adjusting the primary chain

1 To check and adjust the primary chain tension, remove the slotted inspection cap from the primary chaincase outer cover. It is now possible to insert a finger and check the tension of the primary chain in the centre of its run, which should be approximately ½ inch. Check with the engine in several different positions, as the chain may not have worn in an even manner.

2 If it is necessary to make adjustments, slacken the bolt that acts as the gearbox pivot. It passes through the lug on the underside of the gearbox and the lower frame. Slacken also the securing nut which also positions the gearbox adjuster.

3 If it is necessary to increase the chain tension, draw the gearbox backwards by slackening off the nut at the forward facing end of the adjuster and then tighten the nut at the rear, a little at a time whilst checking the chain tension. Only a little movement is usually necessary. When the adjustment is correct, check with the engine in several different positions, then tighten the forward facing adjuster nut. Tighten the remaining gearbox nuts and re-check the tension.

4 To relax chain tension, slacken the rearward facing nut and tighten the nut that is forward facing. The identical procedure should be followed as described in the preceding paragraph.

5 Always re-check after all the gearbox bolts have been tightened. A chain that is too tight will cause harsh transmission and one that is too loose, transmission snatch.

## 7 Removing and replacing the chaincase outer cover

1 The chaincase cover is secured to the chaincase casting by twelve screws. A paper gasket forms the seal between the two mating surfaces, both of which must be scrupulously clean if the chaincase is to remain oiltight.

2 Before the chaincase cover can be removed, it is necessary to remove the bolt securing the left hand silencer to the frame (if fitted) whilst the oil is draining off, so that the exhaust system clears the cover. Remove the left hand footrest. There is no necessity to remove the rear brake pedal because this can be depressed whilst the cover is lifted clear.

3 Check that the cover is not distorted before coating the jointing surface with gasket cement. If the distortion is only slight, the chaincase cover can be rubbed down until it is flat. Coat the jointing face of the chaincase casting with which the cover mates.

4 Position the paper gasket on the cover and replace all screws. If the screw heads are damaged, it is worth renewing them with Allen screws of the same thread and length. These are often available from Triumph dealers as a replacement kit. When all screws and nuts are located correctly and fingertight, final tightening can be completed.

5 If the cover still leaks after the oil content (1/3 pint, SAE 20) has been added, check that it is not leaking from the threads of the lower of the two domed nuts and running down the casting to give the impression of a leaking gasket. This form of leakage can be cured by placing a fibre washer under each nut before tightening.

6 Check that the circular plug in the centre of the chaincase is tight.

7 Do not omit to change the oil in the chaincase at regular intervals. Condensation will cause the oil to emulsify and if this occurs, corrosion and rapid wear will take place throughout the primary transmission system. Condensation can be detected by the characteristic rust stains that appear around the chain links.

8 There is a bleed screw in the rear of the primary chaincase to lubricate the final drive chain. The adjustment is correct if the chain remains oily but there is no spread of oil on to the walls of the rear tyre. To increase the feed, the screw should be turned outwards; to decrease it, inwards. There is a spring under the screw head, to prevent it from turning or vibrating out. Note that the crankcase breather vents into the chaincase and will help to keep the oil level topped up.

## 8 Fault diagnosis - clutch

| Symptom | Cause | Remedy |
|---|---|---|
| Engine speed increases but machine does not respond | Clutch slip | Check clutch adjustment. If correct, suspect worn linings and/or weak springs. |
| Difficulty in engaging gears. Gear changes jerky and machine creeps forward, even when clutch is withdrawn fully | Clutch drag<br>Clutch plates worn and/or clutch drums<br><br>Clutch assembly loose on mainshaft | Check adjustment for too much play.<br>Check for burrs on clutch plate tongues and indentations in clutch drum grooves.<br>Check tightness of retaining nut. If loose, fit new tab washer and retighten. |
| Operating action stiff | Damaged, trapped or frayed control cable.<br>Cable bends too acute<br>Pushrod bent<br>Spring adjusters too tight | Check cable and replace if necessary. Re-route cable to avoid sharp bends.<br>Replace.<br>Slacken adjusters and check clutch does not slip. |
| Clutch needs frequent adjustment | Rapid wear of pushrod | Leave slack in cable to prevent continual load on pushrod. Renew rod because overheating has softened ends. |
| Harsh transmission | Worn chain and/or sprockets<br>Weak shock absorber spring or loose assembly | Replace.<br>Replace spring; tighten assembly. |
| Transmission surges at low speeds | Worn or damaged shock absorber rubbers | Dismantle clutch shock absorber and renew rubbers. |

# Chapter 4 Fuel system and Lubrication

## Contents

## Specifications

### Carburettor settings

| Triumph model | 5T | T100 | TR5 | 6T | TR6 | T110 | T120 |
|---|---|---|---|---|---|---|---|
| Carburettor make | ----------Amal---------- | | | SU | ----------Amal---------- | | |
| Type | Monobloc 376 ---------- | | | MC2 | Monobloc 376 ---------- | | * |
| Main jet | 200 | 220 | 220 | 0.090 in | 250 | 250 | 240 |
| Needle jet | 0.1065 ---------- | | | — | 0.1065 ---------- | | |
| Needle type | C | ---------- | | M9 | C | ---------- | |
| Needle position | 3rd | ---------- | | — | 3rd | ---------- | |
| Throttle valve | 376/3½ | ---------- | | — | 376/3½ | ---------- | |
| Pilot jet | 30 | 25 | 25 | | 25 | 25 | 25 |

* *twin carburettors, T120 model*

| | |
|---|---|
| Fuel tank capacities ... ... ... ... ... ... ... | 4 gallons (18 litres) all models except Trophy<br>3 gallons (13.5 litres) Trophy models |
| Oil tank capacity ... ... ... ... ... ... ... | 5 pints (2.84 litres) |

## 1 General description

The fuel system comprises a petrol tank from which petrol is fed by gravity to the float chamber(s) of the carburettor(s). Two petrol taps, with built-in gauze filter, are located one each side beneath the rear end of the petrol tank. For normal running the right hand tap alone should be opened except under high speed and racing conditions. The left hand tap is used to provide a reserve supply, when the main contents of the petrol tank are exhausted.

For cold starting the carburettor(s) incorporate an air slide which acts as a choke controlled from a lever on the handlebars.

As soon as the engine has started, the choke can be opened gradually until the engine will accept full air under normal running conditions.

Lubrication is effected by the 'dry sump' principle in which oil from the separate oil tank is delivered by gravity to the mechanical oil pump located within the timing chest. Oil is distributed under pressure from the oil pump through drillings in the crankshaft to the big ends where the oil escapes and is fed by splash to the cylinder walls, ball journal main bearings and the other internal engine parts. Pressure is controlled by a pressure release valve, also within the timing chest. After lubricating the various engine components, the oil falls back into the crankcase, where it is returned to the oil tank by means of the scavenge pump. A bleed-off from the return feed to the oil tank is arranged to lubricate the rocker arms and valve gear, after which it falls by gravity via the pushrod tubes and the tappet blocks, to the crankcase. An additional, positive oil feed is arranged from drillings in the timing cover to lubricate the exhaust tappets. It will be noted that the oil pump is designed so that the scavenge plunger has a greater capacity than the feed plunger, this is necessary to ensure that the crankcase is not flooded with oil, and that any oil drain-back whilst the machine is standing is cleared quickly, immediately the engine starts.

## 2 Petrol tank - removal and replacement

1  The petrol tank is secured to the frame by two studs underneath the nose, one on each side. These studs project through two short brackets welded to the frame and are cushioned by rubber washers to damp out vibration. The tank is retained at the rear by two further bolts, in similar fashion. There may be some variations of this arrangement on later models.

2  When the four bolts are removed and the two fuel pipe unions disconnected at their joint with the petrol taps, the tank can be lifted from the machine. Make sure the shaped rubbers are not lost, since they will be displaced as the tank is removed.

3  When replacing the tank, special care must be taken to ensure none of the carburettor control cables are trapped or bent to a sharp radius. Apart from making control operation much heavier, there is risk that the throttle may stick since there is minimum clearance between the underside of the petrol tank and the top frame tube.

4  As a precaution against the bolts working loose, they can be wired together with locking wire.

## 3 Petrol taps - removal and replacement

1  The petrol taps are threaded into inserts in the rear of the petrol tank, at the underside. Neither tap contains provision for turning on a reserve quantity of fuel. It is customary to use the right hand tap only so that the left hand tap will supply the reserve quantity of fuel, unless the machine is used for high speed work or racing. In these latter cases, it is essential to use both taps in order to obviate the risk of fuel starvation.

2  Before either tap can be unscrewed and removed, the petrol tank must be drained. When the taps are removed each gauze filter, which is an integral part of the tap body, will be exposed.

3  When the taps are replaced, each should have a new sealing washer to prevent leakage from the threaded insert in the bottom of the tank. Do not overtighten; it should be sufficient just to commence compressing the fibre sealing washer.

## 4 Petrol feed pipes - examination

1  Petrol feed pipes of different types have been used, with a union connection to each petrol tap and a push-on fit at the carburettor float chamber.

2  After lengthy service, plastic pipes will discolour and harden gradually due to the action of the petrol. There is no necessity to renew the pipes at this stage unless cracks become apparent or the pipe becomes rigid and brittle.

## 5 Carburettor(s) - removal

1  Both single and twin carburettor fitments have been used depending on the version. Early models used the Standard Type 6 Amal carburettor(s) whilst later versions use either the Amal Monobloc Concentric carburettor(s). All types are described here but special emphasis is given to the Concentric because it is, by now, the most usual fitment or replacement.

2  Before removing a carburettor it is first necessary to detach the mixing chamber top which is retained by two small screws and lift away the top complete with the control cables, throttle valve and air slide assemblies. The petrol pipe can then be pulled off the push connection at the float chamber (or the union complete detached) and, after detaching the two retaining nuts and shakeproof washers, the complete carburettor, may be removed from the cylinder head.

3  Some of the older models will have an Amal Concentric carburettor fitted as this is the only type of Amal carburettor now available. It will replace the earlier types and possibly show benefit in terms of both performance and fuel economy if adjusted correctly.

## 6 Carburettor(s) - dismantling, examination and reassembly

### Amal Concentric carburettor only

1  To remove the float chamber, unscrew the two crosshead screws on the underside of the mixing chamber. The float chamber can then be pulled away complete with float assembly and sealing gasket. Remove the gasket and lift out the horseshoe-shaped float, float needle and spindle on which the float pivots.

2  When the float chamber has been removed, access is available to the main jet, jet holder and needle jet. The main jet threads into the jet holder and should be removed first, from the underside of the mixing chamber. Next unscrew the jet holder which contains the needle jet. The needle jet cannot be removed until the jet holder has been unscrewed and removed from the mixing chamber because it threads into the jet holder from the top. There is no necessity to remove the throttle stop or air adjusting screws.

3  Check the float needle for wear which will be evident in the form of a ridge worn close to the point. Renew the needle if there is any doubt about its condition, otherwise persistent carburettor flooding may occur.

4  The float itself is unlikely to give trouble unless it is punctured and admits petrol. This type of failure will be self-evident and will necessitate renewal of the float.

5  The pivot needle must be straight - check by rolling the needle on a sheet of plate glass.

6  It is important that the gasket between the float chamber and the mixing chamber is in good condition if a petrol tight joint is to be made. If it proves necessary to make a replacement gasket, it must follow the exact shape of the original. A portion of the gasket helps retain the float pivot in its correct location; if the pin rides free it may become displaced and allow the float to rise, causing continual flooding and difficulty in tracing the cause. Use Amal replacements whenever possible.

7  Remove the union at the base of the float chamber and check that the inner nylon filter is clean. All sealing washers must be in good condition.

8  Make sure that the float chamber is clean before replacing the float and float needle assembly. The float needle must engage correctly with the lip formed on the float pivot; it has a groove that must engage with the lip. Check that the sealing gasket is placed OVER the float pivot spindle and the spindle is positioned correctly in its seating.

9  Check that the main jet and needle jet are clean and unobstructed before replacing them in the mixing chamber body. Never use wire or any pointed instrument to clear a blocked jet, otherwise there is risk of enlarging the orifice and changing the carburation. Compressed air provides the best means, using a tyre pump if necessary.

10  Before refitting the float chamber, check that the jet holder and main jet are tight. Do not invert the float chamber, otherwise the inner components will be displaced as the retaining screws are fitted. Each screw should have a spring washer to obviate the risk of slackening.

11  When replacing the carburettor, check the O ring seal in the flange mounting is in good condition. It provides an airtight seal between the carburettor flange and the cylinder head flange to

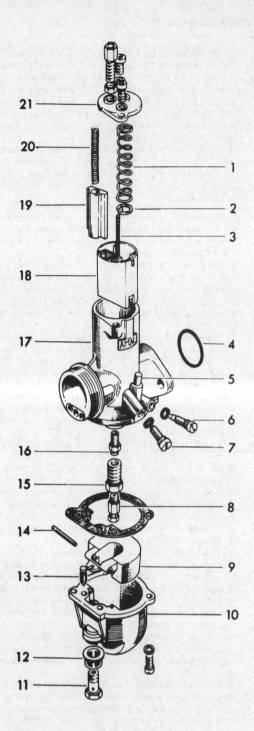

**Fig. 4.1. Amal Concentric carburettor**

| | | | |
|---|---|---|---|
| 1 | Throttle return spring | 11 | Banjo union bolt |
| 2 | Needle clip | 12 | Filter |
| 3 | Needle | 13 | Float needle |
| 4 | 'O' ring | 14 | Float hinge |
| 5 | Tickler | 15 | Jet holder |
| 6 | Pilot jet screw | 16 | Needle jet |
| 7 | Throttle stop screw | 17 | Mixing chamber body |
| 8 | Main jet | 18 | Throttle valve (slide) |
| 9 | Float | 19 | Air slide (choke) |
| 10 | Float chamber | 20 | Air slide return spring |
| | | 21 | Mixing chamber top |

6.22 Early carburettors have a separate float chamber

6.23 Squeeze needle clip to release float

6.23a Needle will withdraw from bottom of float chamber

ensure the mixture strength is constant. Do not overtighten the carburettor retaining nuts for it is only too easy to bow the flange and give rise to air leaks. A bowed flange can be corrected by removing the O ring and rubbing down on a sheet of fine emery cloth wrapped around a sheet of plate glass, using a circular motion. A straight edge will show if the flange is level again, when the O ring can be replaced and the carburettor refitted.

12 Before the mixing chamber top is replaced, check the throttle valve for wear. A worn valve is often responsible for a clicking noise when the throttle is opened and closed. Check that the needle is not bent and that it is held firmly by the clip.

### Amal Monobloc carburettor only

Most models were fitted with the Amal Monobloc carburettor. Since the two designs of carburettor differ in a number of respects, revised procedure is necessary when dismantling, examining and reassembling the Monobloc instrument.

13 The float chamber is an integral part of the Monobloc carburettor and cannot be separated. Access is gained by removing three countersunk screws in the side of the float chamber and removing the end cover and gasket. Remove the small brass distance piece and the float needle which will free from its seating as the float is withdrawn.

14 The main jet threads into the main jet holder which itself is screwed into the main body of the mixing chamber. Removal of the lower main jet cover gives access to the main jet. If the hexagonal nut above the jet cover is unscrewed, the main jet holder can be detached and the needle jet unscrewed from the upper end. The pilot jet has its own separate cover nut. When removed, the jet can be unscrewed. It is threaded at one end and has a screwdriver slot.

15 Unless internal blockages are suspected, or the body is worn badly, there is no necessity to remove the jet block, which is a tight push fit within the mixing chamber body. It is removed by pressing upward, through the orifice of the main jet holder, after removing the small locating peg which threads into the carburettor body. Extreme care must be exercised to prevent distorting either the jet block or the carburettor body which is cast in a zinc-based alloy.

16 Check the float needle for wear by examining it closely. If a ridge has worn around the needle, close to the point, the needle should be discarded and a new one fitted.

17 The float is unlikely to give trouble unless it is punctured, in which case a replacement is essential. Do not omit to fit the small brass distance piece on the float pivot, after the float has been inserted. If this part is lost, there is nothing to prevent the float moving across to the float chamber end cover and binding - a fault that will give rise to intermittent flooding and prove difficult to pinpoint.

18 There must be a good seal between the float chamber end cover and the float chamber. Always use a new gasket when the seal is broken to obviate the risk of continual petrol leakage.

19 Do not omit to inspect and, if necessary, clean the nylon filter within the float chamber union. When replacing the filter, position it so that the gauze is facing the inflow of petrol. On some of the earlier filters, the plastic dividing strips between the gauze segments are somewhat wide and could impede the flow of petrol under full flow conditions.

20 As stressed in the preceding part of this Section, do not use wire or any pointed object to clear blocked jets. Compressed air should be used to clear blockages; even a tyre pump can be utilised if a compressed air line is not available.

21 The Monobloc carburettor has an O ring in the centre of the mounting flange which must be in good condition if air leaks are to be excluded. If the flange is bowed, as the result of previous overtightening, the O ring should be removed and the flange rubbed down on fine emery cloth wrapped around a sheet of plate glass. Run with a rotary motion and when a straight edge shows the flange is level again, the O ring can be replaced.

### Amal type 6 carburettors only

22 The early Type 6 carburettors have a separate float chamber attached to the base of the mixing chamber by a banjo union and retaining nut. If the retaining nut is slackened and removed, the float chamber can be separated, complete with fibre sealing washers.

23 To gain access to the float and float needle assembly of the Type 6 float chamber, unscrew the float chamber lid after slackening the locking bolt close to the outside rim. The float can be lifted out if the needle retaining clip is squeezed free the grip on the needle. The needle itself will drop clear of the base of the float chamber when it is released from the float.

24 Check the float to see whether it has become porous and allows petrol to enter and upset its balance. It should be replaced if a leak is evident. It is not practicable to effect a satisfactory repair.

25 Check the float needle and float needle seating to see whether the float needle is bent or whether a ridge has worn around either the needle or its seating as the result of general wear. All defective parts should be replaced. The needle seat will unscrew from the float chamber body, to permit replacement when necessary.

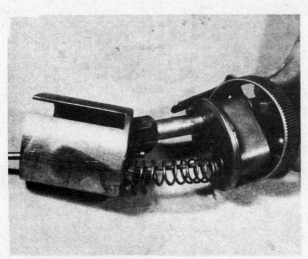

6.25 The throttle valve assembly

6.25a This throttle valve has worn badly - note marks

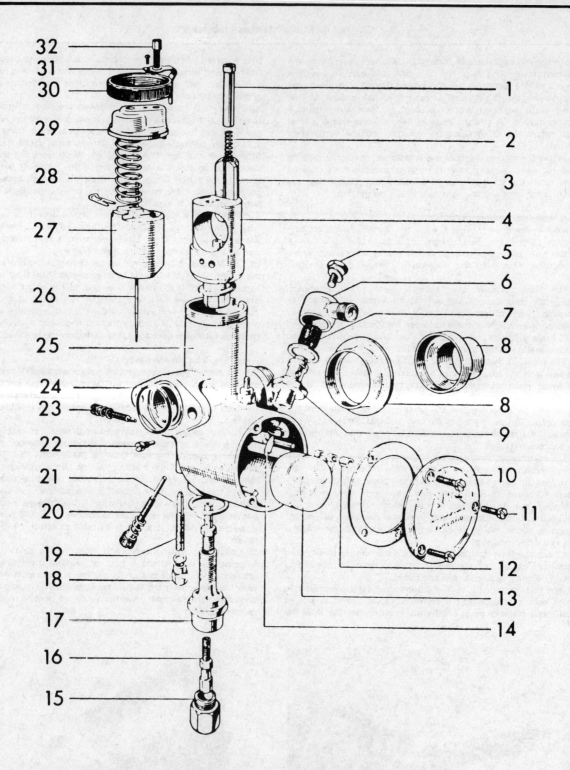

**Fig. 4.2 Amal Monobloc carburettor**

| | | | |
|---|---|---|---|
| 1  Air valve guide | 9  Needle setting | 18  Pilot jet cover nut | 27  Throttle slide |
| 2  Air valve spring | 10  Float chamber cover | 19  Pilot jet | 28  Throttle spring |
| 3  Air valve | 11  Cover screw | 20  Throttle stop screw | 29  Top |
| 4  Jet block | 12  Float spindle bush | 21  Needle jet | 30  Cap |
| 5  Banjo bolt | 13  Float | 22  Locating peg | 31  Click spring |
| 6  Banjo | 14  Float needle | 23  Air screw | 32  Adjuster |
| 7  Filter gauze | 15  Main jet | 24  'O' ring seal | |
| 8  Air filter connection (top) | 16  Main jet | 25  Mixing chamber | |
|    of air intake tube | 17  Main jet holder | 26  Jet needle | |

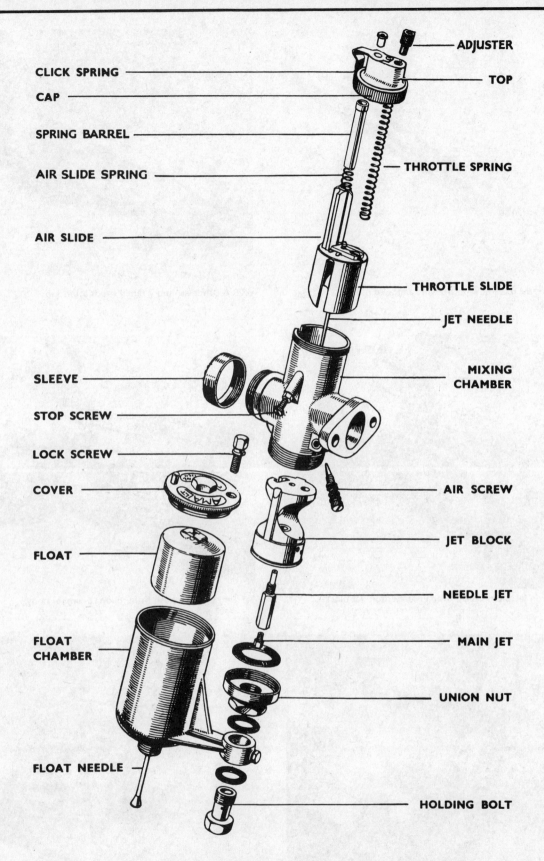

CLICK SPRING

CAP

SPRING BARREL

AIR SLIDE SPRING

AIR SLIDE

SLEEVE

STOP SCREW

LOCK SCREW

COVER

FLOAT

FLOAT CHAMBER

FLOAT NEEDLE

ADJUSTER

TOP

THROTTLE SPRING

THROTTLE SLIDE

JET NEEDLE

MIXING CHAMBER

AIR SCREW

JET BLOCK

NEEDLE JET

MAIN JET

UNION NUT

HOLDING BOLT

Fig. 4.3 Amal type 6 carburettor (early models)

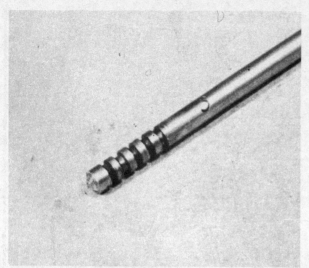

6.26 Needle must be straight. Note engraved number

6.28 Air slide assembly rarely needs attention

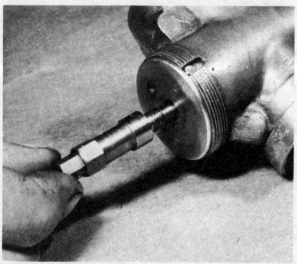

6.29 Main jet and needle jet will unscrew from jet block

7.1 Jet block is a push fit in mixing chamber body

7.2 Note peg, to align jet block correctly

8.4 Throttle valve cutaway is stamped on valve top

26 The throttle slide, needle and air slide assembly still attached to the carburettor top should be examined. Signs of wear on the throttleslide will be self-evident. If the amount of wear is particularly high it may be responsible for a pronounced clicking noise when the engine is running slowly, as the slide moves backwards and forwards within the mixing chamber.

27 The needle should be straight and the needle retaining clip a good fit. Check the needle for straightness by rolling it on a sheet of plate glass. If it is bent, it must be replaced. Reject any needle clip that has lost its tension.

28 The air slide assembly seldom requires attention. Trouble can occur if the compression spring loses its tension, since this will cause the air slide to stick, making cold starting more difficult.

29 Check the main jet, needle jet and pilot jet (if fitted). Wear will occur in the needle jet only; the other jets are liable to blockages if dirty or contaminated petrol is used. NEVER use wire to clear a blocked jet otherwise there is danger of the hole being enlarged. Use either a foot pump or a compressed air line to clear the blockage.

30 On all carburettors there is a pronounced tendency for the mounting flange to 'bow' if the retaining nuts are overtightened. The resultant air leak, which will have a marked effect on carburation, will be difficult to trace as a result. The condition of the flange can be checked by holding a straight edge across the face. If the flange is bowed, it should be rubbed down with a sheet of fine emery cloth wrapped around a piece of flat glass, using a rotary motion, until the bow is removed. Make sure the carburettor body is washed very thoroughly after this operation, to ensure that small particles of abrasive do not lodge in any of the small internal air passages.

## 7  Carburettor(s): removing the jet block(s)

1 The jet block should be a tight push fit in the mixing chamber and it is seldom necessary to disturb it unless it is suspected that some of the small internal airways are blocked.

2 In the Type 6 carburettor, the jet block is pegged to coincide with a slot in the base of the mixing chamber. This ensures correct alignment. To remove the jet block, unscrew the large base nut and withdraw the jet block from the bottom of the mixing chamber.

3 The 'Monobloc' carburettor has a somewhat different arrangement. The jet block will lift out from the top of the carburettor, after the main jet holder has been unscrewed.

4 Do not use force when removing or replacing the jet block. A zinc alloy casting is used for the carburettor body, which is brittle and will distort under certain conditions. Reassemble by reversing the dismantling procedure.

5 The jet block is integral with the Concentric carburettor.

## 8  Carburettor(s) - checking the settings

1 The various sizes of jets and that of the throttle slide, needle and needle jet are predetermined by the manufacturer and should not require modification. Check with the Specifications list if there is any doubt about the values fitted.

2 Slow running is controlled by a combination of the throttle stop and air regulating screw settings. Commence by screwing the throttle stop screw(s) inward so that the engine runs at a fast tickover speed. Adjust the air screw setting(s) until the tickover is even, without either misfiring or 'hunting'. Screw the throttle stop screw(s) outward again until the desired tickover speed is obtained, then recheck with the air regulating screw(s) so that the tickover is as even as possible. Always make these adjustments with the engine at normal running temperature and remember that an engine fitted with high-lift cams is unlikely to run evenly at very low speeds no matter how carefully the adjustments are made.

3 If desired, there is no reason why the throttle stop screw(s) should not be lowered so that the engine will stop completely when the throttle is closed. Some riders prefer this arrangement so that the maximum braking effect of the engine can be utilised on the over-run.

4 As an approximate guide, up to 1/8 throttle is controlled by the pilot jet, from 1/8 to 1/4 throttle by the throttle valve cutaway, from ¼ to ¾ throttle by the needle position and from ¾ to full throttle by the size of the main jet. These are only approximate divisions; there is a certain amount of overlap.

## 9  Balancing twin carburettors

1 Twin carburettors may be fitted, using right and left handed instruments. There may be a balance pipe linking both carburettor inlet ports to improve tickover. A one-into-two throttle cable and air slide cable assemblies is used, each with its own junction box. The junction box components are made of a plastic material; no maintenance is necessary.

2 Before commencing the balancing operation, it is essential to check that both carburettors operate simultaneously. Place a finger inside the bell mouth of each carburettor intake in turn and check when the throttle valve commences to move as the twist grip is rotated. Both slides should begin to rise at exactly the same time; if they do not, use the cable adjusters to ensure the moment of lift coincides. It is important that the throttle stop screws are slackened off during this operation to obviate the risk of a false reading.

3 Cross-check by noting the points at which the throttle slides lift completely and again, adjusting if necessary.

4 Start the engine and when it is at running temperature, stop it and remove one spark plug lead. Restart the engine and adjust the air regulating screw and throttle stop screw of the OPPOSITE cylinder as detailed in Section 2 until the desired tickover speed is obtained. Stop the engine again, replace the spark plug lead and repeat the whole operation with the other cylinder and carburettor.

5 When both spark plug leads are replaced, it is probable that the tickover speed will be too high. It can be reduced to the desired level by unscrewing both throttle stop screws an identical amount and rechecking to ensure both throttle valves still lift simultaneously.

## 10  The SU type MC2 carburettor

### 6T Thunderbird model only

1 The 6T Thunderbird model is the only Triumph twin to which a carburettor of SU manufacture is fitted, as standard. This type of carburettor is based on an entirely different design concept from that of the Amal carburettor, fitted to the other models in the range. In consequence, special instructions for the correct operation and maintenance must be observed.

2 The only item that can be varied is the jet needle, which as standard is the M9 type. Unless carburation problems with a sidecar machine are encountered, the needle should not be changed for one of another specification. It should be noted, however, that some sidecar models perform better when an M7 needle is fitted in place of the original.

3 To gain access to the needle, remove the screws that retain the domed dashpot cover in position, and lift off the cover, taking care to catch the piston and spring within. Great care is necessary when removing the cover as it is only too easy to tilt the piston and bend the needle.

4 The needle is held in the piston by a small screw. The shoulder of the needle must be level with the face of the piston. When the carburettor is dismantled in this fashion, always check for a bent needle.

5 The piston must be free to slide with ease within the domed dashpot cover and must NOT be oiled or greased. It should have a polished surface all over that needs only a wipe with a clean duster. To restore the finish, use metal polish only, NEVER

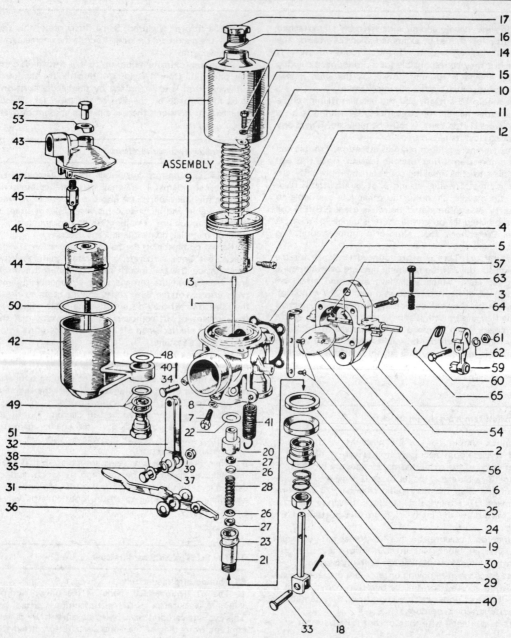

**Fig. 4.4 SU MC2 carburettor (6T model only)**

| | | |
|---|---|---|
| 1 Body | 22 Copper washer | 43 Float chamber lid |
| 2 Throttle barrel adaptor | 23 Copper washer | 44 Float |
| 3 Screw - 4 off | 24 Sealing ring (brass) | 45 Needle and seat |
| 4 Gasket | 25 Sealing ring (cork) | 46 Hinged lever |
| 5 Throttle cable stop | 26 Gland washer (brass) | 47 Hinge pin |
| 6 Screw - 2 off | 27 Gland washer (cork) | 48 Fibre pin |
| 7 Plug screw | 28 Spring | 49 Washers (2 fibre, one brass) |
| 8 Washer | 29 Adjusting nut | 50 Float chamber lid washer |
| 9 Suction chamber complete | 30 Spring | 51 Holding bolt |
| 10 Piston spring | 31 Jet lever | 52 Float chamber lid nut |
| 11 Thrust washer | 32 Jet link | 53 Brass cap |
| 12 Needle screw | 33 Pivot pin (long) | 54 Throttle spindle |
| 13 Jet needle | 34 Pivot pin (short) | 56 Throttle disc |
| 14 Screw - 2 off | 35 Bolt | 57 Screw |
| 15 Spring washer - 2 off | 36 Fibre washer | 59 Throttle lever |
| 16 Oil cap washer | 37 Spring washer | 60 Bolt |
| 17 Oil cap | 38 Washer | 61 Nut |
| 18 Jet | 39 Nut | 62 Washer |
| 19 Jet screw | 40 Split pin | 63 Adjusting screw |
| 20 Jet top half bearing | 41 Return spring | 64 Adjusting screw lock spring |
| 21 Jet bottom half bearing | 42 Float chamber | 65 Lever return spring |

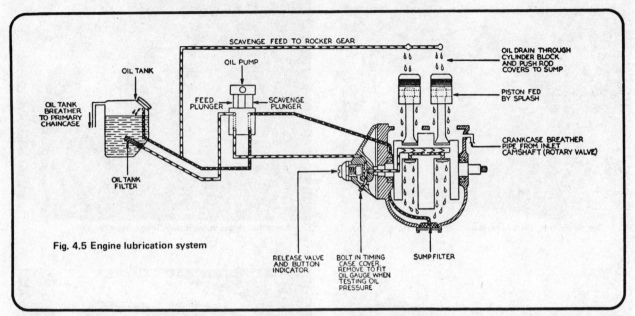

SCAVENGE FEED TO ROCKER GEAR

OIL TANK

OIL PUMP

OIL DRAIN THROUGH
CYLINDER BLOCK
AND PUSH ROD
COVERS TO SUMP

OIL TANK
BREATHER
TO PRIMARY
CHAINCASE

FEED
PLUNGER

SCAVENGE
PLUNGER

PISTON FED
BY SPLASH

CRANKCASE BREATHER
PIPE FROM INLET
CAMSHAFT (ROTARY VALVE)

OIL TANK
FILTER

**Fig. 4.5 Engine lubrication system**

RELEASE VALVE
AND BUTTON
INDICATOR

BOLT IN TIMING
CASE COVER.
REMOVE TO FIT
OIL GAUGE WHEN
TESTING OIL
PRESSURE

SUMP FILTER

EMERY CLOTH. Make sure all traces are removed after the final polish.

6   Mixture strength is established by the nut at the base of the mixing chamber, through which the main jet passes. The main jet can be moved by means of the mixture lever. Once set, further adjustment is rarely necessary. Move the nut only one flat at a time and maintain pressure on the mixture lever so that the jet follows suit. Screwing the nut upwards weakens the mixture and downwards enrichens it. A total of not more than three flats should show a significant difference in engine running. Set for a good, regular engine beat in conjunction with the throttle stop screw. If the setting is correct, the engine will stop if the piston is raised no more than 1/10 inch with a thin rod. If the engine speeds up, the mixture is too rich. During these setting up operations, the engine must be at normal running temperature and the choke open wide.

7   If the engine shows a tendency to stall, will not run slowly or lacks power, it is probable that the piston is sticking in the dashpot. Remove the air filter hose (if fitted) and check that the piston is resting on the bridge piece when the engine has stopped. When raised by hand, it should fall with a distinct click. If it is found to be sticking, remove the dashpot cover and clean and polish the inside, also the outside of the piston itself. When reassembling, note that a groove in the piston locates with an insert in the carburettor body itself.

8   The only maintenance necessary calls for the removal of the cap from the top of the dashpot chamber and the addition of several drops of thin machine oil to the piston rod and guide bush assembly within. This oil performs a damping function for the piston and does not lubricate the main piston assembly. Be sure to replace the plastic cap afterwards, or the air leak that results will upset the carburation.

9   Before suspecting the carburettor itself, make sure the float chamber is not flooding as the result of a damaged float or a badly worn needle valve and seating. If there is any doubt about these parts, renew them as a precaution.

## 11  Air cleaner - removal and replacement

1   Some models are fitted with an air cleaner which takes the form of a separate, circular housing attached direct to the carburettor air intake, or an oblong box with rounded corners, mounted across the frame. Two types of filter element have been employed, a convoluted paper element or one formed from cloth or felt.

2   None of the filter elements should be soaked in oil. It is sufficient to detach the paper element and blow it clean with compressed air or in the case of the cloth or felt elements, to wash them in paraffin and allow them to drain thoroughly before replacing.

3   On no account run the machine with the air cleaner disconnected unless the carburettor has been re-jetted to suit. When an air cleaner is fitted, it is customary to reduce the size of the carburettor main jet, in order to compensate for the enrichening effect of the air cleaner element. In consequence, a permanently-weakened mixture will result if the air cleaner is detached; this will cause failure of the valves and/or piston crown.

## 12  Engine lubrication - removing and replacing the oil pressure release valve

1   Oil pressure is controlled by a pressure release valve located within the timing chest on the right hand side of the engine. Oil pressure, if suspect because the 'tell tale' indicator does not extend, can be verified by attaching an oil gauge to the blanking off plug in the forward-facing edge of the timing cover, using an adaptor. From a cold start, the pressure may rise as high as 80 psi initially but when the engine is at normal running temperature, this reading should drop to 20 or 25 psi at tickover speeds. Normal running pressure should be within the 65 to 80 psi range.

2   If satisfactory readings are not obtained, the engine should be stopped and the pressure release valve dismantled. It is preferable to remove the complete unit by unscrewing the hexagonal nut closest to the timing chest and then dismantle the unit whilst it is clamped in a vice. When the domed cap is removed in this fashion, it will release the spring and the plunger in which the spring seats.

3   Wash the dismantled components in a petrol/paraffin mix and ensure that they are dry before examination. Check that the gauze filter in the end of the unit is not blocked or damaged and that the plunger is free from score marks. The main body of the unit should be unscored internally.

4   Check the free length of the plunger spring. If the length is less than 1 17/32 in the spring must be renewed.

5   Reassemble the unit using new fibre sealing washers. Replace the unit in the timing chest and tighten it fully, with a new fibre washer under the hexagon nut. Start the engine and if the oil

11.1 This air cleaner has a corrugated felt element

12.1 Pressure release valve screws into timing cover

12.3 The fully dismantled release valve assembly

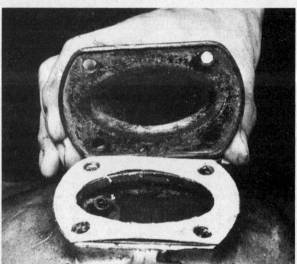

13.1 Use a new gasket when dismantling this filter for cleaning

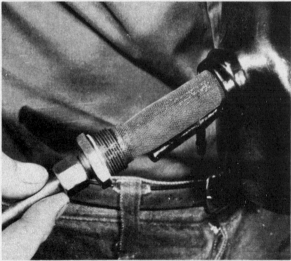

13.3 Oil tank must be drained to gain access to this in-line filter

15.3 Oil pump, showing the feed and scavenge plungers

pressure readings are still low, give attention to the oil pump itself and the filters in the oil tank and the crankcase.

6 Additional parts to check for wear or damage are the rubber sleeve over the indicator button shaft and the tiny O ring seal near the end of the indicator button. Note that the plunger spring is in two parts.

### 13 Engine lubrication - location and examination of oil filters

1 Two filters are included in the engine lubrication system, both of the gauze mesh type. One is located in the underside of the crankcase, blanked off by a cap, accessible from underneath the machine. The other takes the form of an extension of the oil outlet (feed) pipe from the oil tank, within the body of the tank itself.

2 The crankcase filter is withdrawn with the blanking off cap when the latter is unscrewed from the underside of the crankcase. It should be washed with a petrol/paraffin mix and replaced, making sure the cap sealing washer is in good condition.

3 The filter within the oil tank or frame is most unlikely to require attention unless some heavily contaminated oil has been added by accident. To remove the filter, unscrew the hexagon nut above the main feed union. The filter takes the form of a metal gauze, which should be cleansed with petrol and permitted to dry. If the filter is blocked, it will immediately be evident when the main oil feel pipe is detached before the tank is drained.

### 14 Exhaust system - general

1 Most machines have a downswept exhaust system of the two pipe and two silencer type, which may be or may not be joined together by a balance pipe close to the exhaust ports. The Trophy models have twin pipes and silencers upswept on the left hand side of the machine.

2 It is important that the exhaust pipes are a good fit over the stubs which project from the cylinder head. An air leak at this joint will cause the engine to backfire on the over-run.

3 If renewal of either the silencers or exhaust pipes should prove necessary, fit genuine replacement parts of Triumph origin. Although there are many alternative exhaust systems and silencers on the market, it does not necessarily follow that they will improve, let alone maintain, the already high standard of performance. A changed exhaust note is not always indicative of greater speed; in a great many cases the fitting of an exhaust system, not compatible with the engine characteristics, will result in a marked drop in performance.

4 An entirely new type of silencer, is now available that has been evolved to give a significant reduction in noise level without undue power loss. These silencers are recognisable by their long, tapered shape and reverse cone end. They can be fitted as direct replacements to earlier versions.

### 15 Engine lubrication - removal, examination and replacement of oil pump

1 The oil pump is located within the timing cover, which must be removed to gain access. See Chapter 1, section 10. The oil pump is retained by two conical nuts which should not be removed until the oil tank is drained. When both nuts are removed, the oil pump is free to be drawn off the mounting studs.

2 The part subject to most wear is the drive block slider, which should be renewed to maintain pump efficiency. Wear will be obvious immediately on examining the block.

3 The oil pump plungers are not subject to wear because they are continuously immersed in oil. They should, however, be checked for score marks and any undue slackness in the bores. The normal running clearance is up to 0.0005 inch in the case of both the scavenge and feed plungers.

4 Remove the square-headed plugs from the bottom of each plunger housing and check that in each case the ball valve is not sticking to its seat. The springs should have a free length of ½ inch; if they have taken a permanent set they must be renewed.

5 When reassembling the pump, prime both plunger bores with engine oil and check that oil is forced through the outlet ports as the plungers are inserted and depressed. Check that the oil levels within the bores do not fall as the plungers are raised. If they do fall, this denotes a badly seating ball valve at the base of the plunger. Dismantle the ball valve assembly again and clean the seating. The seating can be restored by tapping the ball on its seating with a punch, but only if the pump body is made of brass. Replace the spring and end cap, then recheck again.

6 Always fit a new gasket when the pump is refitted to the timing chest and check that the oil holes align correctly. Do not use gasket cement of any kind, otherwise there is danger of restricting or even blocking the oilways. Check that the drive block silder engages correctly with the ped on the inlet camshaft pinion, before replacing and tightening the conical retaining nuts.

7 Clean off and use new gasket cement on the timing cover when it is replaced to ensure an oiltight joint. Do not forget to replace the blanking off plug when the oil pressure gauge is removed.

8 Remember that most lubrication troubles are caused by failure to change the engine oil at the prescribed intervals. Oil is cheaper than bearings!

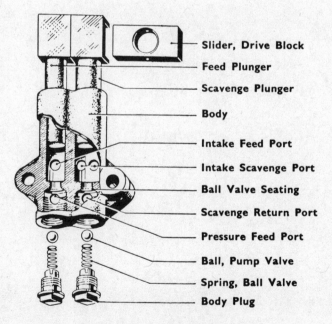

Slider, Drive Block
Feed Plunger
Scavenge Plunger
Body
Intake Feed Port
Intake Scavenge Port
Ball Valve Seating
Scavenge Return Port
Pressure Feed Port
Ball, Pump Valve
Spring, Ball Valve
Body Plug

Fig. 4.6 The oil pump

Fault diagnosis overleaf

## 16 Fault diagnosis - fuel system and lubrication

| Symptom | Cause | Remedy |
|---|---|---|
| Excessive fuel consumption | Air filter choked, damp or oily | Check and if necessary renew. |
| | Fuel leaking from carburettor | Check all unions and gaskets. |
| | Float needle sticking | Float needle seat needs cleaning. |
| | Worn carburettor | Renew. |
| Idling speed too high | Throttle stop screw in too far | Re-adjust screw. |
| | Carburettor top loose | Tighten top. |
| Engine does not respond to throttle | Mixture too rich | Check for displaced or punctured float. |
| | Sticking piston (SU Carburettor) | Remove and clean. |
| Engine dies after running for a short while | Blocked air vent in filler cap | Clean. |
| | Dirt or water in carburettor | Remove and clean float chamber. |
| General lack of performance | Weak mixture; float needle stuck in seat | Remove float chamber and check. |
| | Leak between carburettor and cylinder head | Bowed flange; rub down until flat and replace O ring seal. |
| | Fuel starvation | Turn on both petrol taps for fast road work. |
| | Sticking piston or bent needle (SU carburettor) | Remove, clean and renew damaged part(s). |
| Engine loses power and gets noisy | Lubrication failure | Stop engine immediately and do not re-run until fault is located and remedied. |

# Chapter 5 Ignition system

## Contents

## Specifications

### Ignition timing

Crankshaft degrees/piston position BTDC:

| | |
|---|---|
| 5T 1946 to 1952, 6T 1950 to 1953, TR5 1949 to 1954, T100 ... ... ... ... ... | 37º 3/8 in |
| 5T 1953 to 1959* ... ... ... ... ... ... | 7º 1/32 in |
| T100C ... ... ... ... ... ... ... ... | 42º 15/32 in |
| TR5 1955 to 1959 ... ... ... ... ... ... | 41º 15/32 in |
| 6T 1954* ... ... ... ... ... ... ... | TDC |
| 6T 1955 to 1960* ... ... ... ... ... ... | 6º 1/32 in |
| T110, TR6 ... ... ... ... ... ... ... | 35º 23/64 in |
| T120 ... ... ... ... ... ... ... ... | 39º 7/16 in |

*Note: Ignition timing is set fully advanced on all models except those marked with an asterisk, where it is set fully retarded*

### Contact breaker gap

| | |
|---|---|
| Magneto ignition models ... ... ... ... ... ... ... | 0.012 inch |
| Coil ignition models ... ... ... ... ... ... ... | 0.014 - 0.016 inch |

### Spark plug

| | |
|---|---|
| Champion ... ... ... ... ... ... | L-5/L-7 All cast iron engined models |
| | N-4 Alloy engined models |
| | N-3 T120 models |
| KLG ... ... ... ... ... ... ... ... | FE 80 Alloy engined models |
| Lodge ... ... ... ... ... ... ... ... | 2HLN Alloy engined models |
| NGK ... ... ... ... ... ... ... ... | B7HS Cast iron cylinder heads |
| | B7ES Alloy cylinder heads |
| | B8ES T120 models |
| Spark plug gap ... ... ... ... ... ... | 0.020 inch, magneto ignition models |
| | 0.025 inch, coil ignition models |

### *Magneto

| | |
|---|---|
| Make ... ... ... ... ... ... ... ... | BTH - early models |
| | Lucas - late models |
| Type ... ... ... ... ... ... ... ... | BTH - KC2 |
| | Lucas K2F |
| | Anti-clockwise rotation (drive end) |

### *Ignition coil

| | |
|---|---|
| Make ... ... ... ... ... ... ... ... | Lucas 6 volt |

### *Distributor

| | |
|---|---|
| Make ... ... ... ... ... ... ... ... | Lucas |

*Alternatives

---

### 1  General description

On all but the very late models covered by this manual, the spark necessary to ignite the petrol/air mixture in each combustion chamber is derived from a magneto. The magneto is a self-generating instrument that does not depend on a battery for the initial electrical voltage. This voltage is developed in the primary windings of the armature as it revolves in close proximity to the pole pieces of a magnet. By interrupting the electrical circuit at the appropriate time, using a contact breaker, a very high voltage is developed in the secondary windings of the armature that causes a spark to jump the air gap between the points of the sparking plug.

The magneto is driven from the timing gears and is timed so that each spark will occur at the precise time when it is most necessary to fire the mixture that is under compression in each cylinder. Provision is made to either advance or retard this setting as the engine speed increases or decreases, so that the spark across the points of the sparking plug can be used to maximum advantage.

Because the magneto is a self-generating instrument, the machine can still be used when the battery is detached, or if the electrical system is not in working order, although statutory requirements would make such use illegal. The chief disadvantage of the magneto is that it develops a low voltage at low rotary speeds and in consequence the spark is less intense, which sometimes makes starting more difficult. In practice, such problems rarely occur, provided the magneto is maintained in good working order.

The magneto is now classified as an obsolete instrument, and all production in the UK has ceased. The corresponding shortage of spares makes it necessary to take a defective magneto to a magneto repair specialist, who will either renovate the existing instrument or replace it with a service exchange magneto that has been reclaimed at an earlier date. Magnetos are not necessarily interchangeable because the direction of rotation and the firing angle of the engine have also to be taken into account. On a number of the late type pre-unit construction models, the magneto had been phased out and replaced with an ignition coil and distributor system powered by a crankshaft mounted alternator. An emergency start circuit was included, so that the engine could be started if the battery was fully discharged.

## 2  Removing and replacing the magneto

1  It is not necessary to remove the magneto from the engine unless the engine has to be dismantled completely or the magneto is defective and has to be replaced.

2  Before the magneto can be unbolted from the crankcase, the timing cover must be removed and the magneto drive pinion detached from the magneto drive shaft. If the pinion is fitted with the integral auto-advance unit, it can be detached easily because the retaining nuts act as an extractor. The nut has a normal right-hand thread; it will slacken initially, then commence to tighten as it draws the pinion off the shaft. Continue unscrewing until the pinion pulls clear. It may be necessary to use a sprocket puller on models fitted with manual advance/retard, because the plain driving pinion is bolted direct to the taper of the magneto drive shaft and has no self-extracting action.

3  Before the three nuts retaining the magneto flange to the crankcase are removed, the sparking plug leads should be detached from the sparking plugs (mark each to ensure correct replacement) and the cut-out lead and the advance/retard cable (if fitted). In the latter case it will be necessary to remove the magneto cam ring before the cable can be disengaged.

4  Withdraw the three nuts and their spring washers. The magneto will pull off the flange mounting studs, most probably complete with the sealing gasket that should be discarded.

5  The magneto is replaced by reversing the dismantling procedure. Note that a new sealing gasket must be used to preserve an oil-tight joint at the rear of the timing cover. When the drive pinion is replaced, the ignition must be re-timed by following the procedure detailed in Chapter 1, Sections 30 and 31.

## 3  Magneto - examination and maintenance

1  With the exception of the contact breaker assembly, as discussed in the following section, the only parts of the magneto likely to require attention are the pick-up brushes and the slip ring, which should be checked at regular intervals. There is no necessity to remove the magneto from the machine for this type of maintenance.

2  Remove both pick-up brush holders from the right-hand end of the magneto by moving back the retaining straps. The brush holders will lift out, complete with the pick-up brushes. Access is now available to the slip ring from either housing.

3  The slip ring becomes contaminated with carbon dust from the brushes and oil and grease from the bearings. To clean it, wrap a small piece of rag soaked in petrol around a stick, so that the rag can be pushed on to the brass insert of the slip ring as the magneto is rotated. A pencil is often used for this purpose, with the blunt end downwards, but avoid contact with the point because the lead will act as a pick-up and convey the high tension voltage to the holder! It may be necessary to use a mirror to check when the slip ring is clean once again.

4  Visual inspection will show whether the slip ring is cracked or broken. If either fault is detected, the armature complete with slip ring must be replaced. Water will lodge in a crack or broken portion and cause the high tension voltage to track, leading to a misfire that is very difficult to trace.

5  The brush holders should be cleaned and the brushes checked to ensure they are not too short so that good contact with the slip ring is no longer maintained. Check that each brush holder has a cork sealing gasket to exclude water and oil at the flange joint, and a rubber grommet around the sparking plug lead to form an effective seal at this point.

6  When replacing the brush holders, make sure that both brushes are inserted in their tunnels and do not bind. It is possible for the brushes to slide out of their tunnels if the holders are replaced carelessly, thereby preventing the brushes from contacting the slip ring.

7  If rubber sparking plug leads are fitted, check them for cracks or other surface defects that can cause the spark to jump to a nearby frame or engine component. This form of electrical 'leakage' is most likely to occur when the atmosphere is damp. Plastic-coated cables are recommended, to obviate this trouble.

## 4  Contact breaker - adjustment

1  The contact breaker assembly is located behind the detachable end cover at the extreme left-hand end of the magneto. Adjustment is correct if the gap between the contact breaker points is 0.012" when they are fully open. Check the gap with both cams of the cam ring; if there is a variation the cam ring has worn unevenly and must be replaced.

2  Before checking or adjusting the points gap, examine the surfaces of both points whilst they are in the fully open position. If they are pitted and burnt it will be necessary to remove them for further attention, as described in the following section of this Chapter.

3  Adjustment is effected on the early models by unscrewing the locknut of the fixed contact point and either raising or lowering the point until a 0.012" feeler gauge is a good sliding fit between the fixed and the moveable points. It is essential that the points are in the fully open position whilst the gap is checked or reset. Later models have a different type of contact breaker assembly in which a screw is used to vary the points gap.

4  Before replacing the end cover, give the cam ring a thin smear of grease, taking care none reaches the surface of the contact breaker points. Check that the air hole in the contact breaker end cover is unobstructed, so that the assembly can 'breath'.

## 5  Contact breaker points - removal, renovation and replacement

1  If the contact breaker points are burnt, pitted or badly worn, they should be removed for dressing. If, however, it is necessary to remove a substantial amount of material before the faces can be restored, they should be renewed.

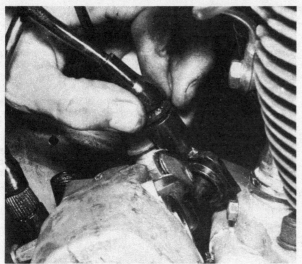

3.2 Remove both pick up brush holders

3.3 Clean slip ring to remove carbon dust and grease

4.1 Check points gap at regular intervals

4.3 Adjust by slackening locknut, then turning fixed contact

5.2 Contact breaker assembly will lift off as a complete unit

5.3 This minute but vital insulator is very easily lost

2 To remove the contact breaker points, first detach the complete contact breaker assembly, which is held on the extreme end of the armature by a centre bolt. Withdraw the bolt and pull the contact breaker assembly off the shaft. Note that the underside of the contact breaker plate has a built-in key, which registers with a corresponding cutaway in the armature shaft. This will ensure the contact breaker is replaced in exactly the same position, thereby eliminating the need to retime the engine.

3 Move the spring arm off the moving contact breaker point (rocker arm) noting that a small but vital insulating washer is interposed between the end of the pivot and the spring arm. Remove this, and place it so that it cannot be lost. Remove the small screw that anchors the return spring of the moving contact to the pillar that forms part of the contact breaker plate. The rocker arm complete with spring can now be lifted off the pivot.

4 The fixed contact breaker point is removed by unscrewing the locknut, then unscrewing the contact point itself.

5 The points should be dressed with an oilstone or fine emery cloth. Keep them absolutely square during the dressing operation, or they will make angular contact when reassembled and will quickly burn away.

6 Replace the contact breaker points by reversing the dismantling procedure. Take particular care when replacing the insulating washers, to ensure they are replaced in the correct order. If this precaution is not observed, the points will be isolated electrically and the magneto will not function. Do not omit the small washer between the moving contact rocker arm and the spring retaining clip.

7 The rocker arm pivot is hollow in many cases and contains a lubricating wick. Add two drops of thin oil to the wick, to ensure the pivot is lubricated. If the contact breaker assembly does not have this facility, apply a thin smear of grease before the rocker arm is replaced.

8 The contact breaker cam ring has an oil hole in the inner face and this too should be lubricated with two drops of thin oil. A thin smear of grease around the inner surface of the ring will help prevent wear of the rocker arm fibre pad.

9 Before the machine is used, check that both contact breaker points have an oil and grease free surface. Quite small electrical currents pass between the points and oil or grease will form an effective insulator.

10 It is advisable to clean the contact breaker assembly every 5000 miles and to check the points gap every 2000 miles.

## 6 Condenser - faults and location

1 A condenser is included in the circuitry of the contact breaker assembly to prevent the points from arcing when they separate. If for any reason the condenser fails, arcing across the points will commence, giving them the characteristic burnt appearance. The most noticeable effect will be a persistant engine misfire, because the spark at the sparking plugs will be less intense. The machine will also be much more difficult to start.

2 The condenser is located within the armature. If the condenser is at fault, it is beyond the means of the average rider to effect a satisfactory repair because the armature will have to be stripped and rewound. The type of repair should be entrusted to a magneto repair specialist.

## 7 Magneto cut-out

1 Most multi-cylinder machines are provided with a means of isolating the magneto electrically, as a convenient means of stopping the engine. If the centre retaining bolt of the contact breaker assembly is earthed, the primary circuit is interrupted and a spark will no longer occur because no high tension voltage is developed all the time the primary circuit is earthed.

2 Most Triumph twins have a spring contact within the magneto end cover that houses a small carbon brush. The carbon brush bears directly on the centre bolt of the contact breaker assembly. A lead from the terminal on the end cover connects with a horn-type push button on the handlebars, which earths the current via the handlebars, when the button is depressed.

3 A few models have a variation of the above arrangement, whereby the cut-out button is mounted directly on the outside of the magneto end cover. The operating principle is, however, the same.

4 If a complete ignition failure occurs, it is possible that unintentional earthing of the cut-out may be responsible. It is easy to check by temporarily removing the lead from the magneto end cover to see whether the ignition circuit is restored. Do not run without the cut-out connected. It is the only practical way of stopping the engine quickly if an emergency should arise, such as a sticking throttle.

## 8 Magneto overhauls and repairs

1 Although a magneto will normally give long, reliable service, occasions will occur when repairs are necessary such as the replacement of the armature bearings, the drive shaft oil seal or even an armature rewind. Because the magneto is now classified as an obsolete instrument, spare parts are becoming increasingly difficult to obtain. In consequence, the average rider has a little option other than to entrust any such repair work to a magneto repair specialist.

2 Even if the parts required can be located, it is questionable whether the average rider will have sufficient experience to effect a satisfactory repair. Special precautions have to be observed to prevent the magnet from losing its magnetism when the armature is removed. For these and other reasons, it is suggested that all such repair work should be carried out by a recognised magneto repair specialist, or alternatively the faulty magneto replaced with another of identical type. Note that the direction of drive is always described as viewed from the drive end. Some repair specialists may offer a service exchange system.

## 9 Ignition timing - checking and resetting

1 If the ignition timing is correct, the contact breaker points will be about to separate when the piston in the cylinder about to fire is the specified distance before top dead centre on the compression stroke.

This method of timing will involve the use of a timing stick, inserted through one of the spark plug holes, to measure the distance of the piston from top dead centre. Although this method is sufficiently accurate for normal road use, optimum engine performance is obtained by using a degree disc attached to the end of the crankshaft. In this case there is a slight variation in the individual settings, as follows:

2 Irrespective of the method of timing used, always turn the engine backwards beyond the intended setting first, then slowly advance it forwards, to the desired setting. This will eliminate any backlash in the timing pinions that may give an inaccurate setting. It is also important that the contact breaker gap is checked, and if necessary, re-set first.

3 Full details of the ignition timing procedure, for both the magneto and coil ignition engines, is given in Chapter 1, Sections 30 and 31 respectively.

4 Optimum performance depends on very accurate setting of the ignition timing. Even a small error in the setting can have a marked effect on both engine performance and petrol consumption. Before the ignition is checked or re-set, it is

7.2 End cover has carbon brush for cut-out connection

9.1 Timing can be checked with vernier, if cylinder head is off

9.7 Cam ring must be in fully advanced position, whilst timing engine

imperative that the contact breaker gap is first set correctly. Changes made after the timing has been set will affect the accuracy of the setting.

5 No advantage is gained from over-advancing the ignition timing in the hope that greater speed will result. Over-advanced ignition timing will make the machine more difficult to start and it will not run evenly at low speeds. In an extreme case it will cause pre-ignition, which will rapidly destroy the crowns of both pistons.

6 Retarded ignition timing is just as harmful. If the ignition is retarded the engine will overheat and will perform in a very sluggish manner. Petrol consumption will rise appreciably and starting may become more difficult.

7 Note that the ignition is usually timed with the ignition control fully advanced. It is impracticable to time the engine with the ignition fully retarded because the different forms of ignition advance do not have an identical range.

## 10 Spark plugs - checking and resetting the gap

1 14 mm spark plugs are fitted to all Triumph twin cylinder models, irrespective of whether the cylinder head is cast in iron or aluminium alloy. Refer to the Specifications section for the list of recommended grades.

2 Models fitted with a cast iron cylinder head use sparking plugs with ½'' reach; ¾'' reach plugs are required for all models fitted with an aluminium alloy cylinder head. Always use the grade of plug recommended, or the direct equivalent in another manufacturer's range.

3 Check the gap at the plug points every 2000 miles. To reset the gap, bend the outer electrode closer to the central electrode and check that a 0.018'' feeler gauge can be inserted. Never bend the central electrode, otherwise the insulator will crack, causing engine damage if particles fall in whilst the engine is running.

4 The condition of the spark plug electrodes and insulator can be used as a reliable guide to engine operating conditions, with some experience. See accompanying illustrations.

5 Always carry at least one spare spark plug of the correct grade. This will serve as a get-you-home means if one of the sparking plugs in the engine should fail.

6 Never overtighten a spark plug, otherwise there is risk of stripping the threads from the cylinder head, especially in the case of one cast in light alloy. A stripped thread can be repaired by using what is known as a 'Helicoil' thread insert, a low cost service of cylinder head reclaimation that is operated by many dealers.

7 Use a spark plug spanner that is a good fit, otherwise the spanner may slip and break the insulator. The plug should be tightened sufficiently to seat firmly on its sealing washer.

8 Make sure the plug insulating caps are a good fit and free from cracks. The caps contain the supressors that eliminate radio and TV interference; in rare cases the suppressors have developed a very high resistance as they have aged, cutting down the spark intensity and giving rise to ignition problems.

## 11 Auto-advance unit

1 Models having a more sporting Specification have manually-operated ignition advance. Other models such as the Speed Twin and Thunderbird have automatic control of ignition advance and retard, obviating the need for the rider to make adjustments in the setting by means of a handlebar control.

2 Standard models have automatic timing control incorporated in the pinion attached to the magneto drive shaft. The pinion carries a plate with two pins, on each of which a weight pivots. The movement of each weight is controlled by a spring located between the end of each pivot weight and a toggle lever close to the centre of each weight. Holes in each toggle lever engage with pegs on the underside of the driving plate, which itself is

connected direct to the magneto spindle. The driving plate is fitted with limit stops, to determine the range of control available. When the magneto is stationary, the weights rest in the closed position and the ignition is retarded fully. When the engine starts, centrifugal force overcomes the restraining action of the springs and the weights move outward, causing movement between the driving pinion and the magneto spindle to occur, which in turn advances the timing. The state of the ignition control therefore remains closely in step with engine requirements, without need to adjust a handlebar control lever manually.

3  The auto-advance unit does not require attention. It is lubricated along with the other pinions in the timing chest and will not need to be disturbed unless it malfunctions or if the timing has to be reset. In the former instance, replacement is necessary because it is impracticable to effect a satisfactory, permanent repair. If the unit is removed during any retiming operation, it must be held in the fully-advanced position whilst the engine timing takes place.

4  On the late-type coil ignition models, the advance-retard unit forms an integral part of the distributor assembly. It operates on an identical principle.

## 12 Distributor - general maintenance

### Alternator models only
1  The distributor is a self-contained unit bolted to the rear of the timing cover in the place originally occupied by the magneto. It rarely requires any attention other than a check to ensure the clamp bolt is tight and the addition of an occasional drop of oil in the centre of the rotor arm spindle, to ensure the contact breaker cam is lubricated. Remove the distributor cover and the rotor arm, and add oil only sparingly. If it gets on to the points, an electrical failure will occur.

2  The distributor must have a waterproof cover fitted, which is not split or torn. If this precaution is ignored, water will eventually work into the unit and cause tracking, with a resultant misfire or even a complete electrical breakdown.

3  If the bushes wear so that the rotor shaft becomes a slack fit, the distributor must be replaced as a complete unit; it is not repairable. If the machine is run with a worn distributor, performance will be affected because the contact breaker gap will constantly vary, changing the ignition timing.

4  The contact breaker assembly is similar in many respects to that of the magneto fitted to the earlier models. When correctly set, the points gap should be within the range 0.014 - 0.016 inch, fully open. The fixed contact point can be moved in relation to the base plate for adjustment purposes, by slackening the retaining screws. Always re-check the gap after the screws have been retightened again.

5  Ignition advance and retard is accomplished by rotating the complete distributor in the desired direction. Full details of the ignition timing procedure are given in Chapter 1, Section 31.

## 13 Fault diagnosis - Ignition system

| Symptom | Cause | Remedy |
|---|---|---|
| Engine will not start | No spark at plugs | Check whether magneto is rotating and whether points open and close. Check also whether points are dirty - if so, clean. |
| | | Check whether points arc when engine is turned over. If so, magneto condenser has failed. Replace armature. |
| | | Cut-out earthed. Remove lead and check whether plugs spark. |
| | | Check whether contact breaker points are opening and also whether they are clean. Check wiring for break or short circuit. |
| Engine starts, but runs erratically | Intermittent or weak spark | Try replacing plugs. |
| | | Check ignition timing. |
| | | Check plug leads for short circuits. |
| Engine will not run at low speeds, kicks back during starting | Ignition over-advanced | Re-set ignition timing. |
| Engine lacks power, overheats | Ignition timing retarded | Re-set ignition timing. |
| Engine misfires at high speeds | Incorrect sparking plugs | Check with list of recommendations |
| Engine fires on one cylinder (coil ignition models only) | No spark at plug or defective cylinder | Check as above, then test ignition coil. If no spark, see whether points arc when separated. If so, renew condenser. |

**Electrode gap check** - use a wire type gauge for best results

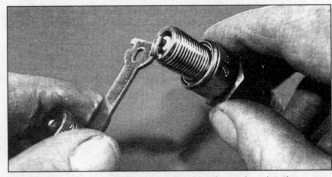

**Electrode gap adjustment** - bend the side electrode using the correct tool

**Normal condition** - A brown, tan or grey firing end indicates that the engine is in good condition and that the plug type is correct

**Ash deposits** - Light brown deposits encrusted on the electrodes and insulator, leading to misfire and hesitation. Caused by excessive amounts of oil in the combustion chamber or poor quality fuel/oil

**Carbon fouling** - Dry, black sooty deposits leading to misfire and weak spark. Caused by an over-rich fuel/air mixture, faulty choke operation or blocked air filter

**Oil fouling** - Wet oily deposits leading to misfire and weak spark. Caused by oil leakage past piston rings or valve guides (4-stroke engine), or excess lubricant (2-stroke engine)

**Overheating** - A blistered white insulator and glazed electrodes. Caused by ignition system fault, incorrect fuel, or cooling system fault

**Worn plug** - Worn electrodes will cause poor starting in damp or cold weather and will also waste fuel

# Chapter 6 Frame and forks

## Contents

## Specifications

### Front Forks

Oil content (per leg):

| | |
|---|---|
| All 1947 - 1959 models ... ... ... ... ... ... | 1/6 pint (95 cc) SAE 20 |
| All 1960 - 1962 solo models ... ... ... ... ... | 1/4 pint (142 cc) SAE 20W/30 |
| All 1960 - 1962 models with longer lower fork legs (sidecar) ... ... ... ... ... ... ... | 3/8 pint (213 cc) SAE 20W/30 |

| Fork springs: | Free length | Colour code |
|---|---|---|
| All 1947 - 1959 models: | | |
| Solo ... ... ... ... ... ... ... | 19.25 in (488.95 mm) | Red |
| Sidecar ... ... ... ... ... ... | 20.00 in (508.00 mm) | Blue |
| Extra-heavy sidecar ... ... ... ... ... | Not available | Purple |
| All 1960 - 1962 models: | | |
| 6T, T110 - solo ... ... ... ... ... ... | 18.31 in (465.07 mm) | Black |
| 6T, T110 - sidecar ... ... ... ... ... | 18.31 in (465.07 mm) | Red/white |
| TR6, T120 - solo ... ... ... ... ... ... | 19.06 in (484.12 mm) | Black/white |
| TR6, T120 - sidecar ... ... ... ... ... | 19.06 in (484.12 mm) | Black/red |

## 1 General description

The very early post-war models covered by this manual have rigid frames, without any means of rear suspension. In due course the Triumph spring hub wheel became available as an optional extra, a unique method of mounting the rear wheel spindle in such a way that it is spring loaded. If this wheel is fitted in place of the standard rear wheel, a limited amount of rear suspension is given to the otherwise rigid frame. The major disadvantage is in the greatly increased weight of the wheel itself.

Ultimately, Triumph Engineering had to modify their frame design in accordance with current practice and revert to the more conventional form of swinging arm rear suspension.

Front fork design remained basically unchanged. The most noticeable visual departure from established practice was the fitting of a nacelle that is integral with the top of the fork unit, in which the switchgear and instruments are mounted in their most convenient position. It was first introduced on the Thunderbird model and was quickly applied to the remainder of the range, with the exception of the Trophy competition models.

## 2 Front forks - removal from frame

1 It is unlikely that the front forks will need to be removed from the frame as a complete unit unless the steering head bearings require attention or the forks are damaged in an accident.

2 Commence operations by placing the machine on the centre stand or front wheel stand (early rigid frame models) and disconnecting the front brake. When using the front stand, the machine must be stood on the rear stand first, or it will topple over. Remove the split pin and clevis pin (or nut and bolt) to disconnect the cable from the operating arm, then unscrew the

2.4 Drain plugs are at base of lower fork legs

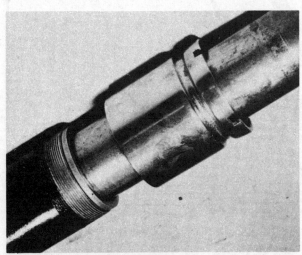

3.2 Unscrew the dust excluder sleeve

3.2a The sealing arrangement on early models

cable adjuster (early models) or pull the cable clear of the stop and the mudguard cable guide (later models). Where appropriate, remove the nut and bolt securing the brake plate to the fork leg.

3   On early models remove the spindle nut, slacken the pinch bolt at the bottom of the left-hand fork leg and tap out the spindle, taking care not to damage the threads. On later models remove the two bolts securing each spindle cap to the fork legs, remove the caps and withdraw the wheel.

4   It is convenient at this stage to drain the fork legs of their oil contents, if the fork legs are to be dismantled at a later stage. The drain plug is found above the wheel spindle recess on each fork leg. Remove both drain plugs and leave the forks to drain into some suitable receptacle whilst the dismantling continues.

5   There is no necessity to remove the front mudguard unless the fork legs are to be dismantled. The lower mudguard stays bolt direct to lugs at the lower end of each fork leg; the centre fixing is made to the inside of each fork leg where a shaped lug accepts the cut-out of the mudguard stays assembly, which is then retained by a bolt and washer. If the various nuts, bolts and washers are removed, the mudguard can be withdrawn, complete with stays.

6   Detach the headlamp after disconnecting the battery leads. (In the case of the Trophy models, this is simplified by the use of a plug-in connector at the base of the shell). Commence by slackening the screw at the top of the headlamp rim, which will allow the rim and reflector unit to be removed. Detach the pilot bulb holder and pull the connectors from the main bulb holder. Disconnect the connectors or the wiring harness and withdraw the wiring harness complete from the headlamp shell. This does not apply to the Trophy plug-in headlamp, which can be removed complete, as a self-contained unit.

7   Remove the pivot bolts on each side of the headlamp shell and withdraw the shell, complete with spacers. On nacelle models, remove the retaining screw at the top of the headlamp rim and ease the headlamp unit away. Disconnect the leads at the snap connectors. Remove the two small screws and nuts that hold the rear of the nacelle unit to the fork shrouds. Unscrew the speedometer cable and disconnect both horn leads.

8   Detach the control cables from the handlebars controls, or the controls themselves, complete with cables. Remove the dipswitch, the cut-out button (if fitted) and the horn push. The handlebars can now be removed by unscrwing the split clamps.

9   Remove the steering damper knob and rod (if fitted) by unscrewing the knob until the rod is released from the lower end into which it threads. Slacken the pinch bolt through the top fork yoke, found at the rear of the steering assembly, above the tank. Unscrew the fork stem sleeve nut (if steering damper fitted) or the blind nut (domed) at the top of the steering head column. Slacken and remove the two plated nuts at the top of each fork leg that pass through the top fork yoke.

10 Remove the top fork yoke by tapping on the underside with a rawhide mallet. The forks should be supported throughout this operation because they will free immediately the yoke clears the tapers of the fork inner tubes or stanchions. When the yoke is displaced and removed, the complete fork assembly can be lowered from the steering head and drawn clear. Note that as the head races separate the uncaged ball bearings will be released. Arrangements should be made to catch the ball bearings as they drop free; most probably only those from the lower race will be displaced.

11 It is possible to remove the fork legs separately, if there is no reason to disturb either the steering head assembly or the fork yokes. In this case the plated top fork nuts should be removed and the pinch bolts through each side of the lower fork yoke slackened and removed. Triumph service tool (Z19) should then be threaded into the top of each stanchion to the full depth of thread and used as a drift to drive each stanchion taper free, to release the stanchion complete with lower leg, so that it can be passed through the lower fork yoke and withdrawn from the machine. It is often necessary to open up the pinch bolt joint in the lower fork yoke to prevent the stanchion from binding. If the stanchion has rusted, this will impede its progress through the bottom yoke. Remove all surface rust with emery cloth,

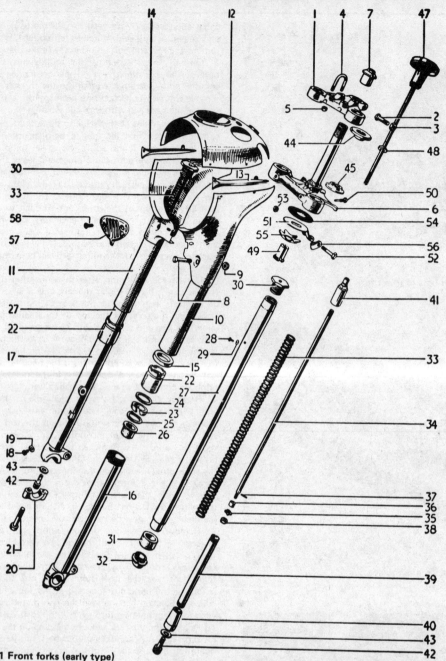

**Fig. 6.1 Front forks (early type)**

1 Upper yoke
2 Pinch bolt
3 Nut
4 Handlebar clamp - 2 off
5 Nut - 4 off
6 Lower yoke and head steam
7 Sleeve nut
8 Pinch bolt - 2 off
9 Stop nut - 2 off
10 Left-hand nacelle cover
11 Right-hand nacelle cover
12 Nacelle top
13 Left-hand motif
14 Right-hand motif
15 Rubber washer - 2 off
16 Left-hand lower fork leg
17 Right-hand lower fork leg
18 Drain plug - 2 off
19 Washer - 2 off

20 Wheel spindle cap - 2 off
21 Spindle cap bolt - 4 off
22 Dust excluder sleeve - 2 off
23 Felt washer - 2 off
24 Washer - 2 off
25 Washer - 2 off
26 Upper bush - 2 off
27 Stanchion - 2 off
28 Oil filler plug - 2 off
29 Washer - 2 off
30 Cap nut - 2 off
31 Lower bush - 2 off
32 Hydraulic stop nut - 2 off
33 Fork spring - 2 off
34 Oil restrictor rod - 2 off
35 Oil restrictor - 2 off
36 Cup - 2 off
37 Cup pin - 2 off
38 Nut - 2 off

39 Pressure tube - 2 off
40 Pressure tube body - 2 off
41 Pressure tube sleeve - 2 off
42 Socket screw - 2 off
43 Aluminium washer - 2 off
44 Top cone and dust cover
45 Lower cone
47 Steering damper knob and rod
48 Damper washer
49 Sleeve
50 Securing pin
51 Damper anchor plate
52 Anchor plate bolt
53 Nut
54 Friction disc
55 Friction plate
56 Speedometer cable clip
57 Horn grille
58 Screw

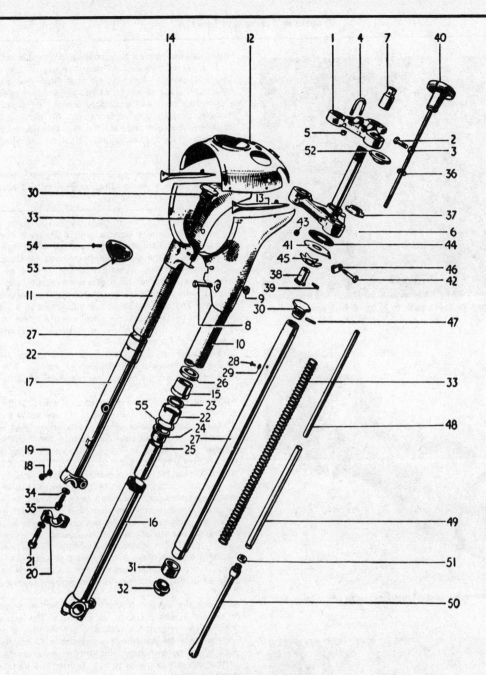

**Fig. 6.2 Front forks (late type)**

 1 Upper yoke
 2 Pinch bolt
 3 Nut
 4 Handlebar clamp - 2 off
 5 Nut - 4 off
 6 Lower yoke and head steam
 7 Head stem nut
 8 Pinch bolt - 2 off
 9 Nut - 2 off
10 Left-hand nacelle cover
11 Right-hand nacelle cover
12 Nacelle top
13 Left-hand motif
14 Right-hand motif
15 Oil seal sleeve - 2 off
16 Lower left-hand fork leg
17 Lower right-hand fork leg
18 Drain plug - 2 off
19 Fibre washer - 2 off

20 Spindle cap - 2 off
21 Spindle cap bolt - 4 off
22 Dust excluder.sleeve - 2 off
23 Oil seal - 2 off
24 Upper bush - 2 off
25 Damping sleeve - 2 off
26 Rubber washer - 2 off
27 Stanchion - 2 off
28 Filler plug - 2 off
29 Fibre washer - 2 off
30 Cap nut - 2 off
31 Lower bush - 2 off
32 Retaining nut - 2 off
33 Fork spring - 2 off
34 Aluminium washer - 2 off
35 Flanged bolt - 2 off
36 Steel washer - 2 off
37 Bottom cone
38 Damper sleeve

39 Locating bolt
40 Steering damper knob and rod
41 Damper anchor plate
42 Bolt
43 Nut
44 Friction disc
45 Star washer
46 Speedometer cable clip
47 Tube guide pin - 2 off
48 Upper guide tube - 2 off
49 Lower guide tube - 2 off
50 Restrictor rod - 2 off
51 Spring support washer - 2 off
52 Upper cone and dust cover
53 Horn grille
54 Screw
55 Washer

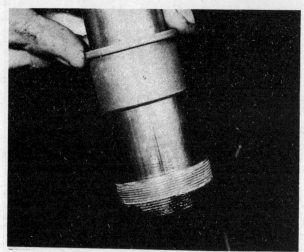

3.3 The upper fork bush is a push fit in lower fork leg

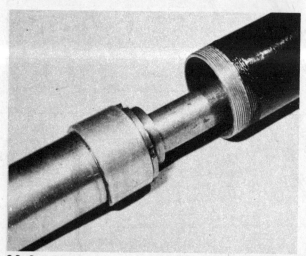

3.3a Stanchion assembly and oil restrictor until will withdraw

3.4 Oil restrictor rod assembly is bolted to end of lower fork leg

wipe clean and apply a light coating of grease or oil.

12 It must be emphasised that the use of the recommended Triumph service tool is essential for this operation. The stanchions are a very tight fit and any attempt to free them with a punch or drift will invariably damage or distort the internal threads, necessitating replacement. An old top fork nut can sometimes be used successfully as a substitute but only if the stanchion is not a particularly tight fit, in either of the fork yokes.

## 3  Front forks - dismantling, examining and reassembling the fork legs

1  When the legs have been removed from the machine, withdraw the nacelle bottom covers. Remove the cork seating washers. Remove the spring abutments and the fork springs.

2  Removal of the dust excluder sleeves which contain a plain washer and oil seal is facilitated by the use of the Triumph service tool. If the service tool is not available, a strap spanner or careful work with a centre punch will provide an alternative solution. The fork leg should be supported in a vice during this operation, by clamping the wheel spindle recess. A sharp knock is needed to free the sleeve initially; thereafter it can be unscrewed and removed.

3  Withdraw the stanchion from each leg by withdrawing it whilst the fork leg is still clamped in the vice. A few sharp pulls may be necessary to release the top bush which is a tight fit in the fork leg.

4  Design changes have occurred in the arrangement of the fork damper units and the following notes will act as a general guide.

5  Some damper units employ an oil restrictor rod assembly, secured by a bolt counter bored into the front wheel spindle recesses of the lower fork legs. Unlike the designs that followed, it is not usually necessary to detach this assembly before the stanchions can be withdrawn from the lower legs.

6  The parts most liable to become damaged in an accident are the fork stanchions, which will bend on heavy impact. To check for misalignment, roll the stanchion on a sheet of plate glass, when any irregularity will be obvious immediately. It is possible to straighten a stanchion that has bowed not more than 5/32 in out of true but it is debatable whether this action is desirable. Accident damage often overstresses a component and because it is not possible to determine whether the part being examined has suffered in this way, it would seem prudent to renew rather than repair.

7  Check the top and bottom fork yokes which may also twist or distort in the event of an accident. The top yoke can be checked by temporarily replacing the stanchions and checking whether they lay parallel to one another. Check the lower fork yoke in the same manner, this time with the stanchions inserted until about 6½ inches protrude. Tighten the pinch bolts before checking whether the stanchions are parallel with one another. The lower yoke is made of a malleable material and can be straightened without difficulty or undue risk of fracture.

8  It is possible for the lower fork legs to twist and this can be checked by inserting a dummy wheel spindle made from 11/16 inch diameter bar and replacing the split retaining clamps. If a set square is used to check whether the fork leg is perpendicular to the wheel spindle, any error is readily detected. Renewal of the lower fork leg is necessary if the check shows misalignment.

9  The fork bushes can be checked by positioning the top bush close to the lower bush at the bottom of the stanchion and inserting the assembly in the lower fork leg. Any undue play will immediately be evident, necessitating renewal of the bushes. If the forks are fitted with the older type of grey sintered iron bushes they should be replaced by the later sintered bronze bushes, to gain the benefit of a reduced rate of wear.

10 Examine both fork springs and check that they are of the same length and have not compressed. If either spring has settled to a length of 1/2 in (12.7 mm) or more shorter than specified, both springs must be renewed. Both must be of the same colour-coding.

3.4a Unscrew pressure tube body....

3.4b ....for access to oil restrictor assembly

3.7 Check yokes for distortion or misalignment

3.9 Lower fork bush is retained by hydraulic stop nut

3.9a After removing nut, bush will pull off

4.3 Use grease to retain ball bearings

### 4  Front forks - examining the steering head races

1  If the steering head races have been dismantled, it is advisable to examine them prior to reassembling the forks. Wear is usually evident in the form of indentations in the hardened cups and cones, around the ball track. Check that the cups are a tight fit in the steering column headlug.

2  If it is necessary to renew the cups and cones, use a drift to displace the cups by locating with their inner edge. Before inserting the replacements, clean the bore of the headlug. The replacement cups should be drifted into position with a soft metal drift or even a wooden block. To prevent misalignment, make sure that the cups enter the headlug bore squarely. The lower cone can be levered off the bottom fork yoke with tyre levers; the upper cone is within the top fork yoke and can be drifted out. Clean up any burrs before the new replacements are fitted. A length of tubing which will fit over the head stem can be used to drive the lower cone into position so that it seats squarely.

3  When the cups and cones are replaced, discard the original ball bearings and fit a new set. It is false economy to re-use the originals in view of the very low renewal cost. Models up to 1955 use 22 3/16 in diameter balls in the top race and 20 1/4 in diameter balls in the bottom race; later models use 20 1/4 in diameter balls in each race. Note that when the bearing is assembled, the race is not completely full. There should always be space for one bearing, to prevent the bearings from skidding on one another and wearing more rapidly. Use thick grease to retain the ball bearings in place whilst the forks are being offered up.

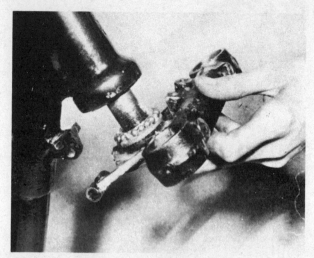

4.3a Leave space for one ball bearing in each race

### 5  Front forks - reassembling the fork legs

1  The fork legs are reassembled by following the dismantling procedure in reverse. Make sure all of the moving parts are lubricated before they are assembled. Fit new oil seals, regardless of the condition of the originals.

2  Before refitting the fork stanchions, make sure the external surfaces are clean and free from rust. This will make fitting into the fork yokes at a later stage much easier. Oil, or lightly grease, the outer surfaces after removing all traces of the emery cloth or other cleaner used.

3  When refitting the fork gaiters (if originally fitted) check that they are positioned correctly. The small hole near the area where the clip fastener is located should be at the bottom.

6.2 Tighten sleeve nut with care, when adjusting head bearings

### 6  Front forks - refitting to frame

1  If it has been necessary to remove the fork assembly complete from the frame, refitting is accomplished by following the dismantling procedure in reverse. Check that none of the ball bearings are displaced whilst the steering head stem is passed through the headlug; it has been known for a displaced ball to fall into the headlug and wear a deep groove around the headstem of the lower fork yoke.

2  Take particular care when adjusting the steering head bearings. The blind or sleeve nut should be tightened sufficiently to remove all play from the steering head bearings and no more. Check for play by pulling and pushing on the fork ends and make sure the handlebars swing easily when given a light tap on one end.

3  It is possible to overtighten the steering head bearings and place a loading of several tones on them, whilst the handlebars appear to turn without difficulty. On the road, overtight head bearings cause the steering to develop a slow roll at low speeds.

4  Before the plated top fork nuts are replaced, do not omit to replace the drain plug in each fork leg and to refill each leg with the correct quantity of SAE 20 Oil.

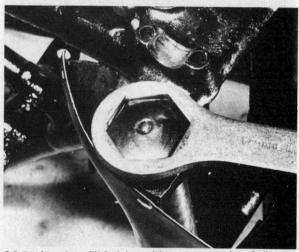

6.3 Don't omit to fill fork legs with correct quantity of oil

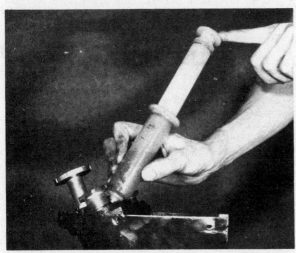

6.3a A syringe makes refilling easy

5   Difficulty will be experienced in raising the fork stanchions so that their end taper engages with the taper inside the top fork yoke. Triumph service tool Z161 is specified for this purpose; if the service tool is not available, a wooden broom handle screwed into the inner threads of the fork stanchion can often be used to good effect.

6   Before final tightening, bounce the forks several times so that the various components will bed down into their normal working locations. This same procedure can be used if the forks are twisted, but not damaged, as the result of an accident. Always retighten working from the bottom upwards.

## 7   Frame assembly - examination and renovation

1   If the machine is stripped for a complete overhaul, this affords a good opportunity to inspect the frame for cracks or other damage which may have occurred in service. Check the front down tube immediately below the steering head and the top tube immediately behind the steering head, the two points where fractures are most likely to occur. The straightness of the tubes concerned will show whether the machine has been involved in a previous accident.

2   If the frame is broken or bent, professional attention is required. Repairs of this nature should be entrusted to a competent repair specialist, who will have available all the necessary jigs and mandrels to preserve correct alignment. Repair work of this nature can prove expensive and it is always worthwhile checking whether a good replacement frame of identical type can be obtained from a breaker.

3   The part most likely to wear during service is the pivot and bush assembly of the swinging arm rear fork. Wear can be detected by pulling and pushing the fork sideways, when any play will immediately be evident because it is greatly magnified at the fork end. A worn pivot bearing will give the machine imprecise handling qualities which will be most noticeable when traversing uneven surfaces.

4   On rigid frame machines, there is no part of the frame that will be subjected to normal wear and tear. A check for cracks and on the overall alignment should prove sufficient.

## 8   Swinging arm rear suspension - examination and renovation

1   If wear is evident in the swinging arm pivot, it will be necessary to remove the swinging arm fork. Commence by placing the machine on the centre stand and remove both the final drive chain and the rear wheel complete with sprocket, following the procedure detailed in Chapter 7. Remove the rear chainguard.

2   The swinging arm fork is retained by a hollow, ground spindle which is a drive fit into the frame lug and a working fit in the two phosphor bronze bushes pressed into the bridge across the swinging arm fork that acts as the pivot. A rod, held by a nut at one end, passes through the centre of the hollow, ground spindle, to retain it in position and to clamp the dust excluders over the ends. There is a spacer between the swinging arm fork and the frame lug on the right-hand side of the machine only, to control the working clearance of the pivot. This should not exceed 0.015 in. on a new machine. A grease nipple is fitted to the frame so that the assembly can be greased at regular intervals.

3   The bearing bushes can be expected to last for approximately 20,000 miles under average riding conditions, before renewal is necessary.

4   To remove the swinging arm assembly, disconnect the rear suspension units from the swinging fork ends by removing the nuts, bolts and washers securing the lower end of each unit to the fork lugs. Remove the nut and tab washer from the rod which passes through the hollow spindle immediately behind the gearbox and withdraw the hollow spindle by driving it out of the frame lug. This is a difficult task as it is a tight fit and it is preferable to have previous experience of this type of repair as well as the appropriate equipment necessary to extract the spindle without risk of damage. When the spindle has been withdrawn, the swinging fork can now be pulled away from the machine, rearwards. The fork will withdraw complete with the bearings and the flanged washers on each side of the pivot. Use a soft metal drift to displace the bushes, which are a good press fit in the pivot tube.

5   If the new bushes are pressed in carefully, use a smear of grease to aid assembly, the correct working clearance should be achieved without need for reaming. Check that the bushes have no burrs before reassembling them.

6   Before replacing the swinging fork in the frame, check that it is straight and does not have a lateral twist. Pack the assembly with new grease, refit the flanged washers with new internal seals and re-align the fork with the frame lug so that the pivot spindle can be pushed home and the rod tab washer and nut refitted. Tighten the nut and bend the tab washer to lock it in place, then reconnect the ends of the rear suspension units. The remainder of the reassembly work should be accomplished by following the dismantling procedure, in reverse.

## 9   Rear suspension units - examination

1   Only a limited amount of dismantling can be undertaken because the damper unit is an integral part of each unit and is sealed. If the unit leaks oil, or if the damping action is lost, the unit must be replaced as a whole after removing the compression spring and outer shroud.

2   If the units require attention, place the machine on the centre stand and remove both nuts and bolts, so that each unit can be detached from the machine. The spring and outer shroud are removed by clamping the lower end of the unit in a vice and depressing the outer shroud so that the split collets which seat in the top of the shroud can be displaced. The shroud and spring can then be lifted off. Note that the suspension units should be set to their light load position, to make this task easier. The standard rating for the springs is 145 lb in and the fitted length 8 inches. The springs should be colour-coded blue/yellow. Lighter springs are available rated at 100 lb in having a fitted length of 8.4 inches. These springs are colour-coded green/green.

3   When replacing the units, make sure the rubber bushes in each 'eye' are a tight fit and in good condition.

## 10 Rear suspension units - adjusting the loading

1   As mentioned earlier in the text, the units can be adjusted without detaching them from the machine, so that the loading can be adjusted to that most suitable for the conditions under which the machine is to be used. A built-in cam at the lower end of each unit permits the sleeve carrying the lower end of the compression springs to be rotated, so that the sleeve is adjustable to different heights. The lowest position, or lightest loading is recommended for solo riding, the middle or medium loading for a heavy solo rider or one with luggage to carry, and the highest or heaviest loading when a pillion passenger is carried. A special 'C' spanner, provided with the tool kit is used to effect the adjustments.
2   Both units must always be set to an identical rating, otherwise the handling of the machine will be seriously affected.

## 11 Centre stand - examination

### Swinging arm models only
1   The centre stand is attached to lugs on the bottom frame tubes to provide a convenient means of parking the machine or raising either wheel clear of the ground. The stand pivots on bolts through these lugs and is held retracted when not in use by a return spring.
2   The condition of the return spring and the return action of the stand should be checked regularly, also the security of the two nuts and bolts retained by a tab washer. If the stand drops whilst the machine is in motion, it may easily catch in some road obstacle and unseat the rider.
3   Some riders remove the centre stand completely because it is inclined to ground if the machine is cornered vigorously. It is questionable whether such action can be justified because in the event of a puncture, there is no alternative means of supporting the machine whilst either wheel is removed.

### Rigid frame models only
4   These machines, which include the spring hub wheel versions, have a rear stand fitted in lieu of a centre stand, to serve the same function. The rear stand requires only superficial attention, to make sure the two retaining bolts (on which it pivots) are tight, with their locking nuts in position, and that the joining rod between the two 'legs' of the stand has not worked loose or broken.

## 12 Prop stand - examination

1   A prop stand which pivots from the left hand lower frame tube is provided for occasional parking, when it is not considered necessary to use the centre stand.
2   At regular intervals check that the prop stand return spring is in good order and that the pivot nut and bolt are not working loose.

## 13 Footrests and rear brake pedal - examination

1   The footrests are attached to a rod that passes through the rear engine plates. They are prevented from rotating by two small pegs engaging with holes in the chaincase centre on the left-hand side and by pegs integral with the right-hand engine plate. A taper joint on the built-up assembly permits the mounting angle to be adjusted to individual requirements.
2   If the machine is dropped, it is probable that the footrests will bend; they are quite soft. To straighten, they must be removed from the machine and their rubbers detached. They should be held in a vice during the straightening operation, using a blow lamp to heat the area where the bend occurs to a cherry red. If they are bent cold, there is a risk of fracture.

3   The rear brake pedal pivot passes through a frame lug brake pedal and the lever itself engages with the lever carrying the brake rod. If the lever should bend, it can be straightened in a similar manner, after removal from the machine.

## 14 Oil tank - removal and replacement

1   Under normal circumstances, it is unlikely that the oil tank will need to be removed from the machine.
2   Drain the oil tank by removing the drain plug at the base of the tank or by pulling off the flexible main feed pipe, after slackening the retaining clip or unscrewing the union nut. Remove the filler cap, the flexible return feed pipe (after slackening the retaining clip), the flexible connection to the froth tower of filler orifice neck (if fitted) and the flexible connection at the T junction of the short return feed pipe used to form a convenient take-off for the rocker feed.
3   The method of removal will vary, according to the model of machine and the year of manufacture. It will be necessary to remove the battery in most cases, so that access is available to the oil tank mounting lugs. Whilst the tank is removed from the machine, it is advisable to unscrew and clean the main filter and at the same time to clean out the tank itself.
4   Replace the tank by reversing the dismantling procedure. When the flexible oil pipes are replaced, check that the inside bore is not rough or beginning to flake. Under these circumstances, particles of rubber can break away and impede the flow of oil; renewal of the pipe is essential. The clips must be tightened fully, otherwise a pipe may come adrift and spray the rear tyre with oil.
5   Do not omit to replace the drain plug before the tank is refilled with engine oil of the recommended grade.

## 15 Speedometer head - removal and replacement

1   The speedometer head is secured to a bracket that bolts to the fork top yoke or housed in the front fork nacelle. The instrument is held to the bracket by studs which project from the base of the casing.
2   To remove the instrument, detach the drive cable by unscrewing the gland nut where the cable enters the instrument body. Unscrew the bulb holder complete with the bulb used to illuminate the dial and unscrew the two nuts that secure the instrument to the mounting bracket, taking care not to displace the shakeproof washers. The instrument can now be lifted away.
3   Apart from defects in either the drive or the drive cable, a speedometer that malfunctions is difficult to repair. Fit a replacement, or alternatively entrust the repair to an instrument repair specialist, bearing in mind that an efficient speedometer is a statutory requirement. If, in the case of a speedometer, the mileage recordings also cease, it is highly probable that either the drive cable or the drive is at fault and not the speedometer head itself. It is very rare for all recordings to fail simultaneously.

## 16 Speedometer drive cable - examination and renovation

1   It is advisable to detach the speedometer drive cable from time to time, in order to check whether it is adequately lubricated and whether the outer cover is compressed or damaged at any point along its run. A jerky or sluggish movement at the instrument head can often be attributed to a cable fault.
2   To grease the cable, uncouple both ends and withdraw the inner cable. After removing the old grease, clean with a petrol soaked rag and examine the cable for broken strands or other damage.
3   Regrease the cable with high melting point grease, taking care not to grease the last six inches closest to the instrument head. If this precaution is not observed, grease will work into the instrument and immobilise the sensitive movement.

4  If the cable breaks, it is usually possible to renew the inner cable alone, provided the outer cable is not damaged or compressed at any point along its run. Before inserting the new inner cable, it should be greased in accordance with the instructions given in the preceding paragraph. Try and avoid tight bends in the run of a cable because this will accelerate wear and make the instrument movement sluggish.

## 17 Seat - removal

1  Although the early models were fitted with a separate saddle and pillion seat, it is doubtful whether they will have survived. Replacement with the more modern dualseat will have taken place in nearly every case, the dualseat bolting to the lug on the top tube of the frame and the two lugs that previously accepted the ends of the saddle springs. The dualseat also has a third fixing point at the rear, where it joins the rear mudguard stays.
2  The later type of dualseat fitted to the swinging arm models has only two fixing points - at the nose to the same frame lug and at the rear to the upper mounting point of the rear suspension units.

## 18 Petrol tank embellishments - removal

1  Some models have metal tank badges secured one to each side of the tank. Each is held in position by two screws which, when released permits removal of the badge.
2  There is also a chromium plated strip down the middle of the tank that hides the otherwise unsightly seam. The strip hooks around the nose of the tank at the front and is retained at the rear by a crosshead bolt which passes through the rear most tank mounting lug, and is secured by a nut.
3  In accordance with well established Triumph practice, a great number of machines are fitted with a tank top luggage carrier which takes the form of a chromium plated grille. The grille is held to the tank top by four slotted head screws which mate up with tapped inserts in the tank top.
4  Rubber knee grips preserve the paintwork on the sides of the petrol tank. They are retained in place by an adhesive compound.

## 19 Sidecar alignment

1  Using conventional fittings, little difficulty is experienced when attaching a sidecar to any of the Triumph 500 cc or 650 cc twins.
2  Good handling characteristics of the outfit will depend on the accuracy with which the sidecar is aligned. Provided the toe-in and lean-out are within prescribed limits, good handling characteristics should result, leaving scope for other minor adjustments about which opinions vary quite widely.
3  To set the toe-in, check that the front and rear wheels of the motor cycle are correctly in line and adjust the sidecar fittings so that the sidecar wheel is approximately parallel to a line drawn between the front and rear wheels of the machine. Re-adjust the fittings so that the sidecar wheel has a slight toe-in toward the front wheel of the motor cycle, as shown in Fig. 6.6. When the amount of toe-in is correct, the distance 'B' should be from 3/8 to ¾ in less than the distance at 'A'.
4  Lean-out is checked by attaching a plumb line to the handlebars and measuring the distance between 'C' and 'D' as shown in Fig. 6.7. Lean-out is correct when the distance 'C' is approximately 1 inch greater than at 'D'.

## 20 Cleaning the machine - general

1  After removing all surface dirt with a rag or sponge washed frequently in clean water, the application of car polish or wax will give a good finish to the machine. The plated parts should require only a wipe over with a damp rag, followed by polishing with a dry rag. If, however, corrosion has taken place, which may occur when the roads are salted during the winter, a proprietary chrome cleaner can be used.
2  The polished alloy parts will lose their sheen and oxidise slowly if they are not polished regularly. The sparing use of metal polish or a special polish such as Solvol Autosol will restore the original finish with only a few minutes labour.
3  The machine should be wiped over immediately after it has been used in the wet so that it is not garaged under damp conditions which will encourage rusting and corrosion. Make sure to wipe the chain and if necessary re-oil it to prevent water from entering the rollers and causing harshness with an accompanying rapid rate of wear. Remember there is little chance of water entering the control cables if they are lubricated regularly, as recommended in the Routine maintenance Section.

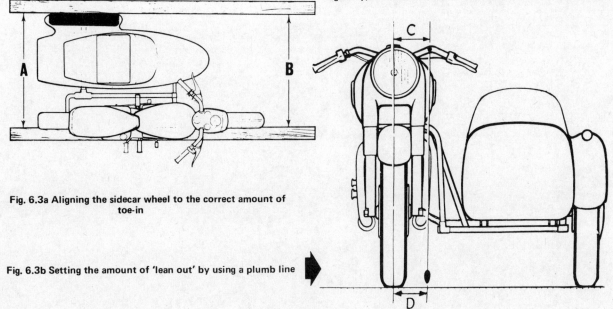

Fig. 6.3a Aligning the sidecar wheel to the correct amount of toe-in

Fig. 6.3b Setting the amount of 'lean out' by using a plumb line

## 21 Fault diagnosis - frame and forks

| Symptom | Cause | Remedy |
| --- | --- | --- |
| Machine is unduly sensitive to road surface irregularities | Fork and/or rear suspension units damping ineffective | Check oil level in forks. Renew suspension units. |
| Machine rolls at low speeds | Steering head bearings overtight or damaged | Slacken bearing adjustment. If no improvement, dismantle and inspect head races. |
| Machine tends to wander. Steering imprecise | Worn swinging arm suspension bearings | Check and if necessary renew bushes. |
| Fork action stiff | Fork legs twisted in yokes or bent | Slacken off wheel spindle clamps, yoke pinch bolts and fork top nuts. Pump forks several times before retightening from bottom. Straighten or renew bent forks. |
| Forks judder when front brake is applied | Worn fork bushes<br>Steering head bearings slack | Strip forks and renew bushes.<br>Re-adjust to take up play. |
| Wheels seem out of alignment | Frame distorted through accident damage | Check frame after stripping out. If bent, specialist repair or renewal is necessary. |
| Machine handles badly under all types of condition | General frame and fork distortion as result of a previous accident or broken frame tube | Strip frame and check for alignment very carefully. If wheel track is out, renew or straighten parts involved. Broken tube will be self-evident. |

# Chapter 7  Wheels brakes and tyres

## Contents

## Specifications

### Wheels

| | |
|---|---|
| Front ... ... ... ... ... ... ... ... | 19 inch diameter, all models except early Trophy TR5 |
| | 20 inch diameter, early Trophy model TR5 |
| Rear ... ... ... ... ... ... ... ... | 19 inch diameter, all models except late Trophy |
| | 18 inch diameter, late Trophy models TR5 and TR6 |

### Brakes

| | |
|---|---|
| Front ... ... ... ... ... ... ... ... | 7 inch diameter, all models except late Tiger 100, T110 and T120 Bonneville |
| | 8 inch diameter, late Tiger 100, T110, TR6 Trophy and T120 Bonneville |
| Rear ... ... ... ... ... ... ... ... | 7 inch diameter, all models |
| | All above brakes are single leading shoe type |

### Tyres

| | |
|---|---|
| Front ... ... ... ... ... ... ... ... | 3.25 x 19 inch, all models except early Trophy TR5 |
| | 3.00 x 20 inch, early Trophy model TR5 |
| Rear ... ... ... ... ... ... ... ... | 3.50 x 19 inch, 5T, Tiger 100, 6T, T110 and T120 |
| | 4.00 x 19 inch, early Trophy model TR5 |
| | 4.00 x 18 inch, late Trophy models TR5 and TR6 |

#### Recommended pressures

| | | |
|---|---|---|
| Front ... ... ... ... ... ... ... ... | 20 psi | 5T, 6T, T110 and T120 models |
| Front ... ... ... ... ... ... ... ... | 20 psi | T100 model |
| Front ... ... ... ... ... ... ... ... | 20 psi | TR5 and TR6 Trophy models |
| Rear ... ... ... ... ... ... ... ... | 20 psi | 5T, 6T, T110 and T120 models |
| Rear ... ... ... ... ... ... ... ... | 20 psi | T100 model |
| Rear ... ... ... ... ... ... ... ... | 18 psi | TR5 and TR6 Trophy models |

The above are solo pressures, based on a rider's weight of 170 lb.

## 1  General description

1  All the models covered by this manual are fitted with 19 inch diameter wheels, with the exception of the Trophy competition models. Early versions of the latter had a 20 inch diameter front wheel to give increased ground clearance and later versions had an 18 inch diameter rear wheel to accommodate the 4 inch section trials tyre fitted as standard.

2  When a machine was supplied to standard specification, the wheels were not of the quickly detachable variety, although on the rigid frame models, the rear mudguard was arranged so that it could be detached very easily to facilitate removal of the rear wheel in the event of a puncture. A quickly detachable rear wheel was, however, available as an optional extra, if ordered with the machine. Both types of wheel are covered in the text of this Chapter.

3  Drum brakes of the single leading shoe type are fitted to both

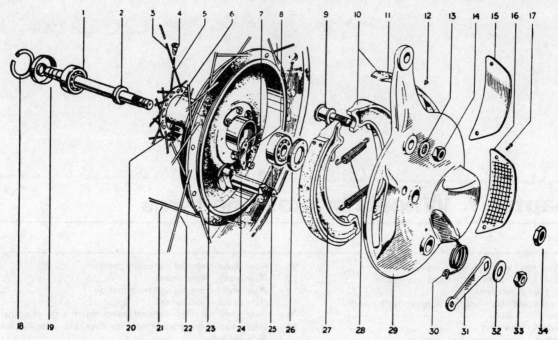

**Fig. 7.1 Front wheel**

1 Wheel bearing
2 Spindle
3 Spoke (90° head)
4 Spoke nipple
5 Spoke (88° head)
6 Hub
7 Lock plate - 4 off
8 Nut - 8 off
9 Brake shoe fulcrum pin
10 Trailing brake shoe
11 Brake shoe lining - 2 off
12 Brake lining rivet - 16 off

13 Washer
14 Fulcrum pin nut
15 Cover plate (alternative)
16 Anchor plate gauze
17 Gauze securing screw - 2 off
18 Bearing circlip (left-hand bearing)
19 Dust cover
20 Brake drum bolt - 8 off
21 Short spoke (95° head)
22 Short spoke (80° head)
23 Brake drum
24 Brake operating cam

25 Wheel bearing
26 Bearing securing ring
27 Leading brake shoe
28 Brake shoe return spring - 2 off
29 Brake plate anchor
30 Brake lever return spring
31 Brake operating lever
32 Washer
33 Brake lever nut
34 Anchor plate nut

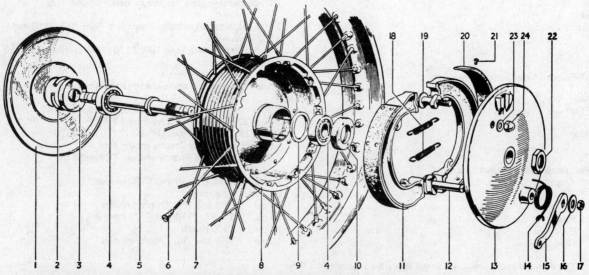

**Fig. 7.2 Front wheel (full width hub)**

1 Cover plate
2 Bearing circlip (left-hand bearing)
3 Dust cover
4 Wheel bearing
5 Spindle
6 Spoke nipple
7 Spoke
8 Hub and brake drum

9 Backing ring
10 Bearing securing ring
11 Brake shoe complete with lining - 2 off
12 Brake operating cam
13 Brake anchor plate
14 Brake lever return spring
15 Brake operating lever
16 Washer

17 Brake lever nut
18 Brake shoe return spring - 2 off
19 Brake shoe fulcrum pin
20 Brake lining - 2 off
21 Brake lining rivet - 16 off
22 Anchor plate nut
23 Washer
24 Fulcrum pin nut

3.1 Remove nut that anchors torque arm to fork leg

3.1a Remove wheel spindle nut, then ....

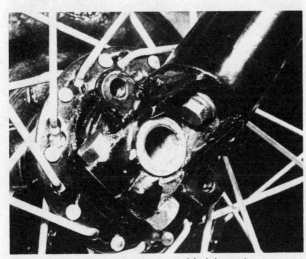

3.1b ....slacken pinch bolt at bottom of fork leg and ....

wheels of the machines described. 7 inch diameter brake drums are fitted as standard, except in the case of the high performance models, where the front brake diameter is increased to 8 inches.

4   An unusual feature on all but the competition models is the fitting of a second, smaller chainguard around the lower run of the final drive chain, to give added protection. This practice ceased with the swinging arm models, when a much deeper main chainguard was fitted to replace the original design.

5   The Triumph spring hub wheel, another optional extra on the early models, is fully covered in a separate Section of this Chapter.

## 2   Front wheel - examination and removal

1   Place the machine on the centre or front stand so that the wheel is raised clear of the ground. Spin the wheel and check for rim alignment. Small irregularities can be corrected by tightening the spokes in the affected area, although a certain amount of skill is necessary if over-correction is to be avoided. Any 'flats' in the wheel rim should be evident at the same time. These are more difficult to remove with any success and in most cases the wheel will have to be rebuilt on a new rim. Apart from the effect on stability, there is greater risk of damage to the tyre bead and walls if the machine is run with a deformed wheel, especially at high speeds.

2   Check for loose or broken spokes. Tapping the spokes is the best guide to the correctness of tension. A loose spoke will produce a quite different note and should be tightened by turning the nipple in an anticlockwise direction. Always check for run-out by spinning the wheel again.

3   If several spokes require retensioning or there is one that is particularly loose, it is advisable to remove the tyre and tube so that the end of each spoke that projects through the nipple after retensioning can be ground off. If this precaution is not taken, the portion of the spokes that projects may chafe the inner tube and cause a puncture.

## 3   Front brake assembly - examination, renovation and reassembly

1   The front brake assembly complete with brake plate can be withdrawn from the front wheel by following the procedure in Chapter 6, Section 2, paragraphs 2 to 3.

2   An anchor plate nut retains the brake plate on the front wheel spindle on later models. When this nut is removed, the brake plate can be drawn away, complete with the brake shoe assembly.

3   Examine the condition of the brake linings. If they are wearing thin or unevenly, the brake shoes should be relined or renewed.

4   To remove the brakes shoes from the brake plate, pull them apart whilst lifting them upward, in the form of V. When they are clear of the brake plate, the return springs can be removed and the shoes separated. Do not lose the pads fitted to the leading edge of each shoe.

5   The brake linings are rivetted to the brake shoes and it is easy to remove the old linings by cutting away the soft metal rivets. If the correct Triumph replacements are purchased, the new linings will be supplied ready-drilled with the correct complement of rivets. Keep the lining tight against the shoe throughout the rivetting operation and make sure the rivets are countersunk well below the lining surface. If workshop facilities and experience suggest it would be preferable to obtain replacement shoes, ready lined, costs can be reduced by making use of a service exchange scheme, available through many Triumph agents.

6   Before replacing the brake shoes, check that the brake operating cam is working smoothly and not binding in its pivot. The cam can be removed for cleaning and greasing by unscrewing the nut on the brake operating arm and drawing the arm off,

3.1c ....withdraw wheel spindle

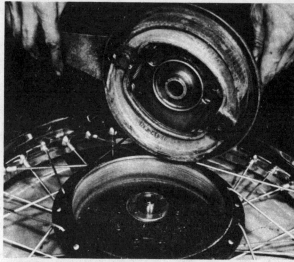

3.2 Brake plate will lift off, with brake shoes

3.4 Lift brake shoes upwards to remove from brake plate

4.1 Bearing retainer has left-hand thread

4.1a Use service tool or pin punch to unscrew

4.1b Left-hand bearing may have retainer or a circlip

4.4 Pack bearings with grease, prior to insertion

4.4a Don't forget the bearing spacer

after its position relative to the cam spindle has been marked so that it is replaced in exactly the same position. The spindle and cam can then be pressed out of the housing in the back of the brake plate.

7   Check the inner surface of the brake drum on which the brake shoes bear. The surface should be smooth and free from score marks or indentations, otherwise reduced braking efficiency is inevitable. Remove all traces of brake lining dust and wipe both the brake drum surface and the brake shoes with a clean rag soaked in petrol, to remove any traces of grease. Check that the brake shoes have chamfered ends to prevent pick-up or grab. Check that the brake shoe return springs are in good order and have not weakened.

8   To reassemble the brake shoes on the brake plate, fit the return springs first and force the shoes apart, holding them in a V formation. If they are now located with the operating cams they can usually be snapped into position by pressing downward. Do not use excessive force or the shoes may distort permanently. Make sure the abutment pads are not omitted.

### 4   Front wheel bearings - removal, examination and replacement

1   On all models withdraw the brake plate. On early models tap out from the brake drum side the sleeve and collar. On all models use a peg spanner or a punch to unscrew **clockwise** (it has a **left-hand** thread) the bearing retaining ring on the hub right-hand side;  note that there may also be a retaining ring on the hub left-hand side. On early models pass a 3/8 in (9.5 mm) diameter drift through the brake drum bearing and tap out the left-hand bearing and dust cap. Withdraw the spacer and tap out to the right the right-hand bearing. On later models remove the circlip from the hub left-hand side and tap the spindle to the left to drive out the left-hand bearing and dust cover, then reverse the spindle and tap out to the right the right-hand bearing, noting the backing ring fitted on full-width hubs only.

2   Wash the bearings in a petrol/paraffin mix to remove all traces of old grease and oil. Clean out the hub and repack it with fresh high melting point grease.

3   When the bearings are dry, check them for play or signs of roughness when they are turned. If there is any doubt about their condition, renew them.

4   On early models, reverse the removal procedure to refit the bearings. Pack both bearings with grease and fill the hub no more than 2/3 full with grease before fitting the second bearing.

5   On later models, pack both bearings with grease. Fit the backing ring (full-width hubs only) and tap the right-hand bearing into the hub until it seats against the hub shoulder.

Tighten (anti-clockwise) the retaining ring. Pack the hub no more than 2/3 full with grease and fit the spindle so that the thread for the brake plate anchor nut is on the right-hand side. Drive the remaining bearing into the hub, followed by the dust cover. Fit the circlip securely into its groove and tap the spindle firmly to the left to press the bearing and dust cover against the circlip. Centralise the spindle before refitting the brake plate.

### 5   Front wheel - replacement

1   Place the front brake assembly in the brake drum and align the front wheel so that the torque anchorage locates with the peg or stud on the lower right hand fork leg. This is most important because the anchorage of the front brake plate is dependent solely on the correct location of these parts. On machines that have the torque arm forming part of the front brake plate, the nut and bolt in the fork leg anchor lug must be replaced and tightened fully. Failure to do so may result in a serious accident, caused by the brake locking on.

2   On early models tighten the spindle nut securely, apply the front brake and pump the forks vigorously up and down to align the fork legs on the spindle. Tighen the pinch bolt securely. On later models the wheel must be positioned so that the grooves in the spindle align with the clamp bolts. Tighten the clamp bolts evenly and securely keeping equal gaps at front and rear between each clamp and its fork leg.

3   Reconnect the front brake cable and check that the brake functions correctly. Check again that the torque arm nut and bolt have been tightened (if fitted).

### 6   Rear wheel - removal and examination

1   Place the machine on the centre or rear stand and before removing the wheel, check for rim alignment, loose or broken spokes and other wheel defects by following the procedure applying to the front wheel, as described in Section 2. On rigid frame models except the TR5, removal of the detachable rear mudguard will help.

2   Two types of rear wheel have been fitted to swinging arm models, the standard or the quickly detachable type. The latter has the advantage of simplified removal, leaving the final drive chain and sprocket in position.

3   If the wheel is of the standard type, commence by disconnecting the final drive chain at the detachable spring link. The task is made easier if the link is first positioned so that it is on the rear wheel sprocket. Unwind the chain off the rear sprocket and lay it on a clean surface.

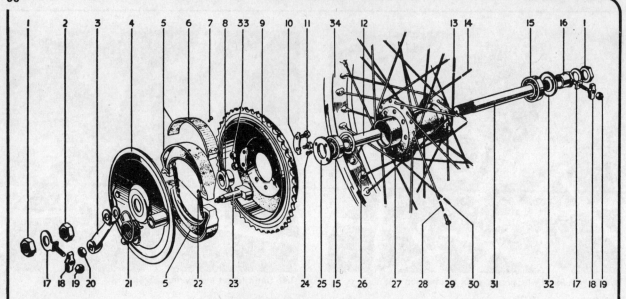

**Fig. 7.3 Rear wheel (standard type, spring frame)**

1 Wheel spindle nut - 2 off
2 Anchor plate locknut
3 Brake operating lever
4 Anchor plate
5 Brake shoe complete with lining - 2 off
6 Brake shoe lining - 2 off
7 Brake lining rivet - 16 off
8 Distance piece, left-hand bearing
9 Brake drum and sprocket
10 Lockplate
11 Brake drum bolt - 8 off
12 Distance piece

13 Spoke (80° head)
14 Spoke (97° head)
15 Wheel bearing - 2 off
16 Right-hand bearing retaining nut
17 Chain adjuster - 2 off
18 Adjuster end plate - 2 off
19 Nut - 2 off
20 Brake operating lever nut
21 Brake lever return spring
22 Brake shoe return spring - 2 off
23 Brake operating cam
24 Bearing retaining ring locking screw

25 Bearing retaining ring
26 Left-hand bearing ring
27 Hub
28 Spoke (76° head)
29 Spoke nipple
30 Spoke (100° head)
31 Wheel spindle
32 Dust cap
33 Distance piece
34 Wheel rim

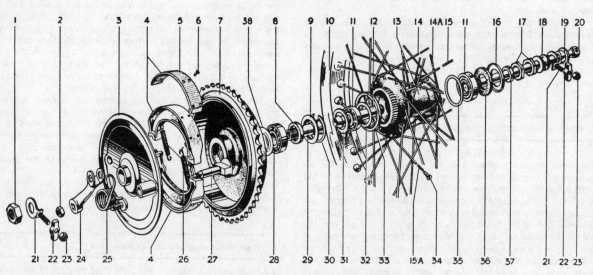

**Fig. 7.4 Rear wheel (quickly-detachable type)**

1 Left-hand wheel nut
2 Brake operating lever nut
3 Brake anchor plate
4 Brake shoe complete with lining - 2 off
5 Brake shoe lining - 2 off
6 Brake lining rivet - 16 off
7 Brake drum and sprocket
8 Felt washer
9 Brake drum bearing retainer
10 Wheel rim
11 Taper roller bearing - 2 off
12 Bearing sleeve
13 Hub

14 Spoke
15 Spoke
16 Dust cap
17 Bearing locknut - 2 off
18 Spindle distance collar
19 Spindle collar
20 Spindle
21 Chain adjuster - 2 off
22 Adjuster end plate - 2 off
23 Nut - 2 off
24 Brake operating lever
25 Brake lever return spring
26 Brake shoe return spring - 2 off

27 Brake operating cam
28 Brake drum bearing
29 Brake drum sleeve
30 Bearing retaining circlip
31 Dust cap
32 Bearing backing ring
33 Hub to brake drum dust seal
34 Spoke nipple
35 Bearing backing ring
36 Felt washer
37 Distance piece (right-hand bearing)
38 Grease retainer

4 Take off the brake rod adjuster and pull the brake rod clear of the brake operating arm. On swinging arm models disconnect the brake torque stay by removing the rear mounting nut and slackening the front mounting nut and bolt.

5 Unscrew both spindle nuts and, on swinging arm models only, raise the rear chainguard by slackening the bottom nut of the left-hand suspension unit. Withdraw the wheel rearward until it drops from the frame ends complete with chain adjusters. It may be necessary to tilt the machine or raise it higher, so that there is sufficient clearance for the wheel to be taken away from the machine.

6 A different procedure is employed in the case of machines fitted with the quickly detachable rear wheel. It is necessary only to unscrew and remove the wheel spindle from the right-hand side of the machine. If the shouldered distance piece between the frame end and the hub is removed, the wheel can be pulled sideways to disengage it from the brake drum centre (splined fitting) before it is lifted away. Note there is a rubber ring seal over the splines which is compressed when the wheel is in position. This acts as a grit seal and must be maintained in a good condition.

### 7 Rear wheel bearings - removal, examination and replacement

1 On rigid frame models, unscrew the spindle nuts and remove the adjuster collars, then unscrew its anchor nut and withdraw the brake backplate. Unscrew the bearing adjuster locknuts and withdraw the spindle. Tap out the bearing inner races and dust caps, passing a drift through the hub from the opposite side, then drive out the outer races and backing rings; renew the backing rings if they are damaged. Check the bearings for wear and renew them if necessary.

2 On reassembly, fit each backing ring and press in each outer race in turn. Pack it with grease and refit the left-hand bearing inner race, then tap the dust cap into place. Turn the wheel over and pack the hub no more than 2/3 full with grease, then pack the right-hand bearing with grease and refit it, followed by its dust cap. Screw the backplate inner locknut (smaller shoulder outwards) to the end of the spindle thread and fit the spindle from the brake drum side. Turn the wheel over, clamp the spindle in a vice, using jaw covers to protect the threads, and re-fit the adjuster locknuts. Tighten the inner locknut hard to settle the bearings, then slacken it fully. Tighten it again until the wheel is free to rotate with just perceptible play at the rim, then tighten the outer locknut hard. Re-check the adjustment, then refit the backplate and second locknut, the adjuster collars and spindle nuts.

3 On swinging arm models with standard rear wheels, unscrew the anchor nut, remove the backplate and spacer, and tap the spindle out to the right. Remove the grub screw locking the bearing retaining ring and use a peg spanner or a punch to remove the ring. *Note: check carefully whether the ring has a left- or right-hand thread before attempting to unscrew it and never*

*use excessive force or the hub may be damaged.* Unscrew the right-hand bearing retaining nut, displace the central spacer to one side, then pass a drift through the hub from the opposite side and tap out each bearing in turn. Withdraw the spacer and backing ring. Check the bearings for wear and renew them if necessary.

4 On reassembly fit its backing ring, pack it with grease and drive the left-hand bearing into the hub then tighten the retaining ring securely; ensure it is turned the correct way. Lock the retaining ring with its grub screw, then turn the wheel over and fit the central spacer. Pack the hub no more than 2/3 full with grease and pack the right-hand bearing with grease before driving it into place. Tap the dust cap into the hub, refit the spindle then turn the wheel over and refit the spacer, backplate and anchor nut. Tighten securely the anchor nut then tighten the right-hand bearing retaining nut hard against the bearing.

5 On swinging arm models with quickly-detachable rear wheels unscrew the two locknuts on the spindle sleeve right-hand end, push the sleeve out from the right and remove the small spacer from the right-hand bearing. Passing a drift through the hub from the opposite side, tap out in turn the inner races and dust caps, followed by the outer races and backing rings. Check the bearings and felt washer for wear or damage and renew them if necessary.

6 On reassembly refit each backing ring and press in each outer race in turn. Pack it with grease and refit the left-hand bearing inner race, then turn the wheel over and pack the hub no more than 2/3 full with grease. Grease the right-hand bearing inner race before refitting it, then insert the spindle sleeve from the brake drum side with its threaded end to the right. Press the left-hand dust cap into the hub then refit the felt washer and right-hand dust cap. Insert the small spacer into the right-hand dust cap then refit the two locknuts. Tighten hard the inner locknut to settle the bearings then slacken it at least 1/4 turn until the sleeve rotates freely and there is just perceptible play at the wheel rim. Tighten the outer locknut on to the inner then recheck the adjustment.

### 8 Rear brake - removal and examination

1 If the rear wheel is of the standard type or if the rear wheel has been removed together with the brake drum and sprocket (quickly detachable wheels) the rear brake assembly is accessible when the brake plate is lifted away from the brake drum.

2 If the quickly detachable wheel has been removed in the recommended way (as described in Section 6.6) it will be necessary to detach the brake drum and sprocket assembly from the frame. This is accomplished by removing the final drive chain by detaching the spring link, unscrewing and removing the nut from the brake plate torque stay so that the latter can be pulled away, and unscrewing and removing the large nut around the bearing sleeve that supports the brake and sprocket assembly. Note that the brake drum has a bearing in its centre which

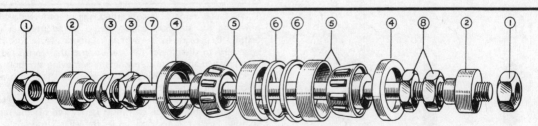

**Fig. 7.5 Rear wheel spindle assembly (rigid frame models only)**

1 Wheel nut - 2 off
2 Chain adjuster thrust collar - 2 off
3 Locknut, rear brake anchor plate - 2 off
4 Bearing dust cap - 2 off
5 Taper roller bearing - 2 off
6 Bearing backing ring - 2 off
7 Rear wheel spindle
8 Bearing adjusting locknut - 2 off

should be knocked out, cleaned and examined before replacement.

3   The rear brake assembly is similar to that of the front wheel drum brake and is of the single leading shoe type with only one operating arm. Use an identical procedure for examining and renovating the brake assembly to that described in Section 3, paragraphs 3 to 6.

## 9   Rear brake - replacement

1   Reverse the dismantling procedure when replacing the front brake to give the correct sequence of operations. In the case of the quickly detachable rear wheel, the brake drum and sprocket should be fitted to the frame first to aid assembly.

2   It is important not to omit the rubber sealing ring of the quickly detachable wheel which fits over the splines of the hub and is compressed when the splines are located with those cut within the brake drum centre. The seal must be in good condition if it is to prevent the entry of road grit and other foreign matter which may cause rapid wear of the splines.

## 10   Rear wheel and gearbox final drive sprockets - examination

1   Before replacing the rear wheel, it is advisable to examine the rear wheel sprocket. A badly worn sprocket will greatly accelerate the rate of wear of the final drive chain and in an extreme case, will even permit the chain to ride over the teeth when the initial drive is taken up. Wear will be self-evident in the form of shallow or hooked teeth, indicating the need for early renewal.

2   In the case of the standard wheel, the sprocket is secured to the brake drum by eight nuts and bolts. When a quickly detachable wheel is fitted, the sprocket is an integral part of the brake drum, in which case the complete unit must be renewed.

3   The gearbox sprocket should also be inspected closely at the same time because it is considered bad practice to renew the one sprocket alone. A certain amount of dismantling work is necessary before the gearbox sprocket can be removed.

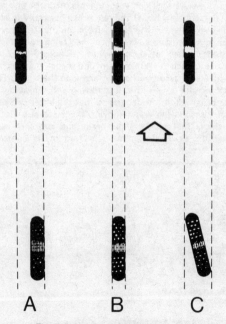

**Fig. 7.6 Checking wheel alignment**

A & C incorrect
B correct

## 11   Rear wheel - replacement

1   When replacing the rear wheel on rigid frame models, make sure the stud on the rear brake plate locates with the channel of the rear left-hand frame end. Failure to observe this precaution will cause the brake to lock on when applied and may give rise to a serious accident.

2   In the case of swinging arm machines make sure the torque arm nut and bolt are replaced and tightened fully.

3   Check to ensure wheel alignment is correct.

## 12   Final drive chain - examination and lubrication

1   The final drive chain is not fully enclosed. The only lubrication provided takes the form of an oil bleed from the primary chaincase that distributes excess oil to the lower chain run.

2   Chain adjustment is correct when there is approximately ¾ inch play in the middle of the chain run, measured at either the top or the bottom. Always check at the tightest spot of the chain run with the rider seated normally.

3   If the chain is too slack, adjust by slackening the wheel spindle and/or wheel nuts and the nut of the torque arm stay, then drawing the wheel rearward by the chain adjusters at the end of each swinging arm fork. It is important to ensure that each adjuster is turned an equal amount so that the rear wheel is kept in alignment. When the correct adjusting point has been reached, push the wheel forward and tighten the wheel nuts and/or spindle, not forgetting the torque arm nut. Recheck the chain tension and wheel alignment, before the final tightening.

4   To check whether the chain needs renewing, lay it lengthwise in a straight line and compress it endwise so that all play is taken up. Anchor one end firmly, then pull endwise in the opposite direction and measure the amount of stretch. If it exceeds ¼ inch per foot, renewal is necessary; Never use an old or worn chain when new sprockets are fitted; it is advisable to renew the chain at the same time so that all new parts run together.

5   Every 2000 miles remove the chain and clean it thoroughly in a bath of paraffin before immersing it in a special chain lubricant such as Linklyfe or Chainguard. These latter types of lubricant are applied in the molten state (the chain is immersed) and therefore achieve much better penetration of the chain links and rollers. Furthermore, the lubricant is less likely to be thrown off when the chain is in motion.

6   When replacing the chain, make sure the spring link is positioned correctly, with the closed end facing the direction of travel. Replacement is made easier if the ends of the chain are pressed into the teeth of the rear wheel sprocket whilst the connecting link is inserted, or a simple 'chain-joiner' is used.

## 13   The Triumph sprung hub wheel - removal, examination and replacement

1   As far as removal is concerned, the procedure described for removing the standard wheel should be adopted, as given in Section 6, paragraphs 3 - 5 of this Chapter. Note that it will be necessary to disconnect the torque arm that forms part of the rear brake plate by unscrewing and removing the bolt and two locknuts that pass through the frame lug. This bolt acts as a pivot as well as the anchorage.

2   Before the brake plate can be freed to gain access to the brake shoes, the torque arm must be levered off the rear wheel spindle. Place a screwdriver under the lever, close to the spindle, and tap the other end with a hammer. There are two split collars on the underside of the lever. Withdraw the dust excluder centre sleeve, spring and the sliding portion.

3   The dust cover is retained by two screws and can be eased off with a screwdriver once these two screws are removed. This will permit the brake plate to be lifted out, complete with the brake assembly.

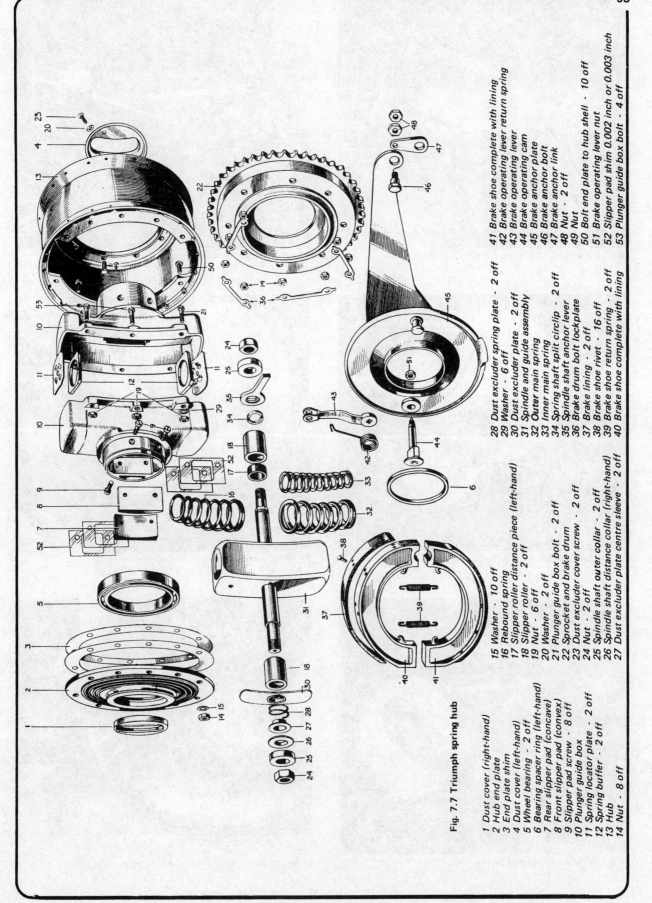

**Fig. 7.7 Triumph spring hub**

1 Dust cover (right-hand)
2 Hub end plate
3 End plate shim
4 Dust cover (left-hand)
5 Wheel bearing - 2 off
6 Bearing spacer ring (left-hand)
7 Rear slipper pad (concave)
8 Front slipper pad (convex)
9 Slipper pad screw - 8 off
10 Plunger guide box
11 Spring locator plate - 2 off
12 Spring buffer - 2 off
13 Hub
14 Nut - 8 off

15 Washer - 10 off
16 Rebound spring
17 Slipper roller distance piece (left-hand)
18 Slipper roller - 2 off
19 Nut - 6 off
20 Washer - 2 off
21 Plunger guide box bolt - 2 off
22 Sprocket and brake drum
23 Dust excluder cover screw - 2 off
24 Nut - 2 off
25 Spindle shaft outer collar - 2 off
26 Spindle shaft distance collar (right-hand)
27 Dust excluder plate centre sleeve - 2 off

28 Dust excluder spring plate - 2 off
29 Washer - 6 off
30 Dust excluder plate - 2 off
31 Spindle and guide assembly
32 Outer main spring
33 Inner main spring
34 Spring shaft split circlip - 2 off
35 Spindle shaft anchor lever
36 Brake drum bolt lockplate
37 Brake lining - 2 off
38 Brake shoe rivet - 16 off
39 Brake shoe return spring - 2 off
40 Brake shoe complete with lining

41 Brake shoe complete with lining
42 Brake operating lever return spring
43 Brake operating lever
44 Brake operating cam
45 Brake anchor plate
46 Brake anchor bolt
47 Brake anchor link
48 Nut - 2 off
49 Nut
50 Bolt end plate to hub shell - 10 off
51 Brake operating lever nut
52 Slipper pad shim 0.002 inch or 0.003 inch
53 Plunger guide box bolt - 4 off

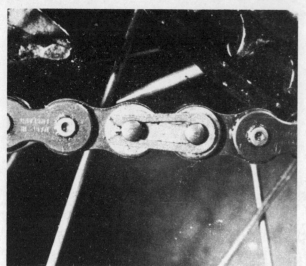

12.6 Ensure chain link faces in the correct direction

13.1 Detach end of torque arm from frame lug

13.2 Slider assembly must be levered out of position

13.2a Note spring and split collars on underside

13.3 Dust cover is retained by two screws

13.3a Torque arm will lift off with integral brake plate

**Tyre changing sequence - tubed tyres**

 Deflate tyre. After pushing tyre beads away from rim flanges push tyre bead into well of rim at point opposite valve. Insert tyre lever adjacent to valve and work bead over edge of rim.

Use two levers to work bead over edge of rim. Note use of rim protectors

 Remove inner tube from tyre

When first bead is clear, remove tyre as shown

 When fitting, partially inflate inner tube and insert in tyre

Work first bead over rim and feed valve through hole in rim. Partially screw on retaining nut to hold valve in place.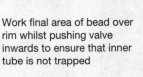

Check that inner tube is positioned correctly and work second bead over rim using tyre levers. Start at a point opposite valve.

Work final area of bead over rim whilst pushing valve inwards to ensure that inner tube is not trapped

13.4 Rear wheel sprocket is retained by eight nuts

13.5 Remove right-hand cover

13.6 Spring box can be withdrawn, after removal of covers

13.6a Note large diameter wheel bearings within covers

13.7 Heed warnings on spring box

13.7a Markings prevent accidental reversal on reassembly

4 Remove the nuts that retain the rear wheel sprocket in position and pull off the sprocket. A few light taps on the right-hand end of the rear wheel spindle should help start it, if it fits tightly on the studs.

5 Turn the wheel over and remove the nuts and washers that retain the right-hand cover in place. If the cover proves difficult to move, tap the left-hand end of the wheel spindle to start it.

6 With both covers removed, the spring box can now be with-drawn from the hub, complete with the two large diameter wheel bearings. These latter bearings are a light drive fit on the arms of the spring box.

7 ON NO ACCOUNT SLACKEN AND REMOVE THE BOLTS THAT HOLD THE TWO HALVES OF THE SPRING BOX TOGETHER. THE SPRINGS WITHIN ARE UNDER GREAT TENSION AND WILL CAUSE SERIOUS INJURY IF THEY ARE RELEASED WITHOUT THE PROPER EQUIPMENT. THIS IS DEFINITELY A JOB FOR THE TRIUMPH SPECIAL-IST — THE AUTHOR HAS SEEN HOLES IN A GARAGE ROOF CAUSED BY UNSKILLED REMOVAL OF THE SPRINGS. The spring box has cast-in warnings and is also marked to ensure it cannot be replaced upside down.

8 To reassemble, reverse the dismantling operations, after first making sure the spring box is positioned correctly. The rear wheel spindle must be in the upper half of the arcuate slot in which it travels, when the wheel is relocated in the frame.

9 Always double check to ensure the anchor arm is correctly located with the lug on the left-hand side of the frame and that the frame locator aligns with the channel cast in the lug that holds the rear wheel spindle. See accompanying illustrations.

## 14 Front brake - adjustment

1 Brake adjustment is effected by the cable adjuster built into the front brake lever, which should be screwed outward to take up slack which develops in the operating cable as the brake shoes wear. Although adjustment is a matter of personal setting, there should never be sufficient slack in the cable to permit the lever to touch the handlebars before the brake is applied fully.

2 Eventually, braking action will be lost because cable adjustment has resulted in a poor angle between the brake operating arms and the direction of pull, causing loss of leverage. This is because the brake linings have now reached the stage where renewal is necessary. To continue beyond this point is unwise. Apart from reduced braking efficiency, the point will eventually be reached where the brake will no longer pull-off effectively, after application.

## 15 Rear brake - adjustment

1 Rear brake adjustment is effected solely by the screwed adjuster at the extreme end of the brake operating rod. It should be screwed inward, to decrease the amount of play at the brake pedal. Always check after making adjustments to ensure that the brake shoes are not binding.

2 Brake adjustment will be necessary when slack in the rear chain is taken up. Because this involves moving the rear wheel backward in the frame, the rear brake adjuster may have to be slackened off a little.

3 After the rear brake has been adjusted, check the stop light action. It may be necessary to re-adjust the point at which the bulb lights by repositioning the clamp around the brake operating rod connected to the operating spring.

## 16 Front wheel - balancing

1 It is customary, on all high performance machines, to balance the front wheel complete with tyre and tube. The out of balance forces which exist are then eliminated and the handling of the machine improved. A wheel, which is badly out of balance produces throughout the steering, a most unpleasant hammering effect at high speeds.

2 One ounce and half ounce balance weights are available which can be slipped over the spokes and engaged with the square section of the spoke nipples. The balance weights are normally positioned diametrically opposite the tyre valve, which is usually responsible for the out of balance factor.

3 When the wheel is spun it will come to rest with the heaviest point downward; balance weights should be added opposite to this point. Add or subtract balance weights until the wheel will rest in ANY position after it has been spun.

4 If balance weights are not available, wire solder wrapped around the spokes, close to the nipples, is an excellent substitute.

5 There is no necessity to balance the rear wheel for normal road use.

## 17 Speedometer drive gearbox - general

1 Models that do not have the speedometer drive taken from the gearbox require an external drive, which is usually taken from a speedometer drive gearbox fitted to the right hand side of the rear wheel. The drive is transmitted from the hub by means of a slotted locking ring which threads into the hub and performs the dual function of retaining the right hand wheel bearing.

2 Provided this gearbox is greased at regular intervals, it is unlikely to require attention during the normal life of the machine.

3 Speedometer drive gearboxes are not necessarily interchangeable, even though they may look similar. If a replacement has to be made, it is advisable to check the Specification. The drive ratio is related to the size of the rear wheel and the section of tyre fitted, two variables which will have a marked effect on the accuracy of the speedometer reading.

4 When the drive is taken direct from the gearbox it is important to change the speedometer drive pinions within the gearbox itself if the overall gear ratios are altered. This will ensure speedometer accuracy is retained.

## 18 Tyres - removal and replacement

1 At some time or other the need will arise to remove and replace the tyres, either as the result of a puncture or because a renewal is required to offset wear. To the inexperienced, tyre changing represents a formidable task yet if a few simple rules are observed and the technique learned the whole operation is surprisingly simple.

2 To remove the tyre from either wheel, first detach the wheel from the machine by following the procedure given in this Chapter whether the front or the rear wheel is involved. Deflate the tyre by removing the valve insert and when it is fully deflated, push the bead of the tyre away from the wheel rim on both sides so that the bead enters the centre well of the rim. Remove the locking cap and push the tyre valve into the tyre.

3 Insert a tyre lever close to the valve and lever the edge of the tyre over the outside of the wheel rim. Very little force should be necessary; if resistance is encountered it is probably due to the fact that the tyre beads have not entered the well of the wheel rim all the way round the tyre.

4 Once the tyre has been edged over the wheel rim, it is easy to work around the wheel rim so that the tyre is completely free on one side. At this stage, the inner tube can be removed.

5 Working from the other side of the wheel, ease the other edge of the tyre over the outside of the wheel rim furthest away. Continue to work around the rim until the tyre is free completely from the rim.

6 If a puncture has necessitated the removal of the tyre, reinflate the inner tube and immerse it in a bowl of water to trace the source of the leak. Mark its position and deflate the tube.

Dry the tube and clean the area around the puncture with a petrol soaked rag. When the surface has dried, apply rubber solution and allow this to dry before removing the backing from a patch and applying the patch to the surface.

7  It is best to use a patch of the self-vulcanising type, which will form a very permanent repair. Note that it may be necessary to remove a protective covering from the top surface of the patch, after it has sealed in position. Inner tubes made from synthetic rubber may require a special type of patch and adhesive if a satisfactory bond is to be achieved.

8  Before replacing the tyre, check the inside to make sure that the agent which caused the puncture is not trapped. Check the outside of the tyre, particularly the tread area, to make sure nothing is trapped that may cause a further puncture.

9  If the inner tube has been patched on a number of past occasions, or if there is a tear or large hole, it is preferable to discard it and fit a new tube. Sudden deflation may cause an accident, particularly if it occurs with the front wheel.

10  To replace the tyre, inflate the inner tube just sufficiently for it to assume a circular shape. Then push it into the tyre so that it is enclosed completely. Lay the tyre on the wheel at an angle and insert the valve through the rim tape and the hole in the wheel rim. Attach the locking cap on the first few threads, sufficient to hold the valve captive in its correct location.

11  Starting at the point furthest from the valve, push the tyre bead over the edge of the wheel rim until it is located in the central well. Continue to work around the tyre in this fashion until the whole of one side of the tyre is on the rim. It may be necessary to use a tyre lever during the final stages.

12  Make sure there is no pull on the tyre valve and again commencing with the area furthest from the valve, ease the other bead of the tyre over the edge of the rim. Finish with the area close to the valve, pushing the valve up into the tyre until the locking cap touches the rim. This will ensure the inner tube is not trapped when the last section of the bead is edged over the rim with a tyre lever.

13  Check that the inner tube is not trapped at any point. Reinflate the inner tube, and check that the tyre is seating correctly around the wall of the tyre on both sides, which should be equidistant from the wheel rim at all points. If the tyre is unevenly located on the rim, try bouncing the wheel when the tyre is at the recommended pressure. It is probable that one of the beads has not pulled clear of the centre well.

14  Always run the tyres at the recommended pressures and never under or over-inflat. See Specifications for recommended pressures.

15  Tyre replacement is aided by dusting the side walls, particularly in the vicinity of the beads, with a liberal coating of French chalk. Washing-up liquid can also be used to good effect, but this has the disadvantage of causing the inner surfaces of the wheel rim to rust.

16  Never replace the inner tube and tyre without the rim tape in position. If this precaution is overlooked there is good chance of the ends of the spoke nipples chafing the inner tube and causing a crop of punctures.

17  Never fit a tyre which has a damaged tread or side walls. Apart from the legal aspects, there is a very great risk of a blow-out, which can have serious consequences on any two-wheel vehicle.

18  Tyre valves rarely give trouble but it is always advisable to check whether the valve itself is leaking before removing the tyre. Do not forget to fit the dust cap which forms an effective second seal. This is especially important on a high performance machine, when centrifugal force can cause the valve insert to retract and the tyre to deflat without warning.

## 19 Security bolt

1  It is often considered necessary to fit a security bolt to the rear wheel of a high performance model because the initial take up of drive may cause the tyre to creep around the wheel rim and tear the valve from the inner tube. The security bolt retains the bead of the tyre to the wheel rim and prevents this occurrence.

2  A security bolt is fitted to the rear wheel of the Triumph competition and certain of the high performance models as a safety precaution. Before attempting to remove or replace the tyre, the security bolt must be slackened off completely.

## 20 Fault diagnosis - wheels, brakes and tyres

| Symptom | Cause | Remedy |
|---------|-------|--------|
| Handlebars oscillate at low speeds | Buckle or flat in wheel rim, most probably front wheel. | Check rim alignment by spinning wheel. Correct by retensioning spokes or rebuilding on new rim. |
| | Tyre not straight on rim | Check tyre alignment. |
| Machine lacks power and accelerates poorly | Brakes binding | Warm brake drum provides best evidence. Re-adjust brakes. |
| Brakes grab when applied gently | Ends of brake shoes not chamfered | Chamfer with file. |
| | Elliptical brake drum | Lightly skim in lathe (specialist attention required). |
| Brake pull-off sluggish | Brake cam binding in housing | Free and grease. |
| | Weak brake shoe springs | Renew if springs have not become displaced. |
| Harsh transmission | Worn or badly adjusted final drive chain | Adjust or renew as necessary. |
| | Hooked or badly worn sprockets | Renew as a pair. |
| | Loose rear sprocket (standard wheel only) | Check sprocket retaining bolts. |

# Chapter 8  Electrical system

## Contents

## Specifications

### Battery
| | |
|---|---|
| Make ... ... ... ... ... ... ... ... ... | Lucas |
| Amp/hr capacity ... ... ... ... ... ... ... | 13 or 22 (sidecar use) |
| Volts ... ... ... ... ... ... ... ... ... | 6 volts, positive earth system |

### *Dynamo
| | |
|---|---|
| Make ... ... ... ... ... ... ... ... ... | Lucas |
| Type ... ... ... ... ... ... ... ... ... | E3L |
| Output (max) ... ... ... ... ... ... ... | 60 watts, d.c. |
| Voltage ... ... ... ... ... ... ... ... | 6 volts |

### *Alternator
| | |
|---|---|
| Make ... ... ... ... ... ... ... ... ... | Lucas |
| Type ... ... ... ... ... ... ... ... ... | RM14 |
| Output ... ... ... ... ... ... ... ... | 60 watts, a.c. |
| Voltage ... ... ... ... ... ... ... ... | 6 volts |

### *Rectifier
| | |
|---|---|
| Make ... ... ... ... ... ... ... ... ... | Lucas |

*late models have an alternator, coil and rectifier in place of the magneto and dynamo*

### Bulbs
| | |
|---|---|
| Headlamp main ... ... ... ... ... ... ... | 30/24 watt pre-focus |
| Headlamp pilot ... ... ... ... ... ... ... | 6 watt bayonet fitting |
| Stop and tail ... ... ... ... ... ... ... | 6/18 watt, offset pins bayonet |
| Speedometer lamp ... ... ... ... ... ... | 2 watt, small bayonet fitting |

## 1  General description

Direct current is generated by the separate dynamo mounted at the front of the crankcase and driven by the timing pinions. It is fed to the battery via a voltage regulator, which keeps the level of output in step with the electrical load. During normal daytime running, only a low output is provided, unless the battery itself is in a discharged state. During the hours of darkness, when the lighting system is in use, the output is correspondingly higher, to balance the load without draining the battery. An indication of the charge rate in amperes, is given by means of the ammeter mounted in the headlamp nacelle or the headlamp shell. The dynamo has an output of 60 watts maximum d.c.

On some of the later models, an alternating current generator, or alternator, is mounted on the end of the crankshaft. The output is used to power the ignition system, and also to charge the battery, after the alternating current has been changed to direct current by means of a rectifier. Output is regulated by switching extra coils into the circuit, as in the case of running during the hours of darkness when the electric demand is correspondingly higher. In the event of a discharged battery, an emergency circuit provides a means of starting the machine.

## 2   Dynamo - checking the output

1   The output from the dynamo can be checked by removing the electrical lead from the end cover and bridging both terminals together, to form one point for the connection of a d.c. voltmeter. The other connection from the meter (negative) should be attached to the dynamo casing or some other convenient earthing point of the machine. Running on open circuit, the dynamo should give a voltmeter reading of 7 - 8 volts, when the engine is started and is running at a fast tickover.

2   If a voltmeter is not available, visual evidence of a charge can be obtained by connecting a headlamp bulb in the same manner as the voltmeter. Use a 12 volt bulb, otherwise the 6 volt bulb will blow, due to the excess voltage developed when the dynamo is running on open circuit. The bulb should glow quite brightly when the engine is started.

3   If there is no evidence of a charge, check that the dynamo armature is rotating. A loose driving pinion has been known to give the illusion of a faulty dynamo.

4   A dynamo is now regarded as an obsolete instrument and there is no longer the possibility of obtaining a service exchange replacement from the manufacturer at moderate cost. Some of the larger dealers still offer a form of service exchange replacement; the alternative is to entrust the repair to a dynamo repair specialist, who will have the facilities for rewinding the armature or the field coil - the two parts of a dynamo most likely to break down.

## 3   Dynamo - lubricating the bearings and general maintenance

1   The Lucas E3L dynamo fitted to most machines has a ball journal bearing at each end of the armature. These bearings are packed with grease and will not normally require attention until the machine itself is due for a complete overhaul. It is the opportune to re-pack the bearings with high melting point grease, provided there is no play or roughness as they are rotated. Always test by dismantling the dynamo and removing the bearings, so that they can be washed out first.

2   The armature should be cleaned periodically, to remove the accumulation of carbon dust from the brushes. Use a rag moistened in petrol, held against the armature as the latter is rotated. If the surface is particularly dirty, very fine sandpaper can be used to clean the soft copper segments. Make sure no traces of abrasive are left behind.

3   The mica insulation between the individual segments of the commutator should lie slightly below the surface. After an extensive period of use, the commutator may wear to such an extent that the mica begins to stand proud. Under these circumstances, the dynamo should be dismantled and the mica slightly undercut, using a small portion of fine hacksaw blade. This is a very delicate operation.

4   The most common cause of reduced or even complete lack of output is worn brushes. The wear limit is about half the normal full length of a new brush. If in doubt, renew the brushes as a precaution. Note that the brushes must slide freely in their holders and that there must be adequate spring pressure to keep them in contact with the commutator.

5   The cork gasket between the end of the dynamo housing and the back of the timing cover must be in good condition if oil leaks are to be avoided. It is permissible to use a light smear of gasket cement when making this joint.

6   To dismantle the dynamo after it has been removed from the machine, remove the clamp around the commutator end and lift out both brushes from their holders. Remove the screw from the centre of the end cover, then lift the end cover away. Unless

necessary, it is convenient to leave the dynamo leads attached. Unscrew the two long screws that pass right through the dynamo body. When the have been withdrawn, complete with their nuts and shakeproof washers, the commutator end can be pulled away from the main body; it is dowelled into position. The other end will pull away complete with the armature; if the armature is to be separated, it will be necessary to first pull the drive pinion off the keyed shaft. The field coil will be left inside the main body and should not be disturbed unless absolutely necessary. If it works loose, the pole piece will hit the revolving armature and cause irreparable damage.

## 4   Voltage regulator - function and adjustments

1   The voltage regulator, which takes the form of a small, oblong box clamped to some convenient part of the machine (or housed in the toolbox), performs a dual function. It contains the cut-out, an electro-magnetic device that determines the point at which the dynamo is connected to the charging circuit and the point at which it is disconnected. If this provision was not included in the electrical system, the battery would discharge through the dynamo when it was stationary or running at low speeds, causing damage to both the battery and the field coil of the dynamo. As its name implies, the regulator also controls the output from the dynamo, when it is connected to the charging circuit, by means of another electromagnetic method. This explains the presence of the two separate coils in the regulator unit, each with its own set of contacts. The regulator matches the charge from the dynamo to the requiremens of the battery, hence if the battery is discharged, a full charge will be given. If, on the other hand, the battery is fully charged, the charge rate will be cut down so that only a trickle charge is given until a heavy load is again placed on the battery.

2   The regulator is correctly adjusted during manufacture and further adjustments should not be required until the machine has seen a considerable amount of service. The parts most likely to require attention are the contacts, which should be cleaned with either fine emery cloth or an oilstone. Complete the operation by cleaning with methylated spirits, to remove all traces of dust and foreign matter.

3   It will probably be necessary to remove the armature plate to gain access to the contact points of the regulator coil, in which case the air gap between the core of the coil and the armature plate will have to be reset. This is accomplished by slackening off the locknut on the voltage adjusting screw and unscrewing the screw until it is clear of the flat spring that tensions the armature plate. Slacken the two screws that secure the armature plate to the main body of the regulator unit and insert a 0.015'' feeler gauge between the armature plate until it is squarely in contact with the feeler gauge and tighten the two retaining screws. Before the feeler gauge is withdrawn turn the voltage adjusting screw until it just touches the armature plate tension spring and tighten the locknut. Check, and if necessary reset the voltage adjusting screw, as described in the following Section.

4   If the cutout points are dismantled for cleaning, they should be reset so that there is a gap of 0.025'' - 0.030'' between the stop arm and the moving contact, when the contact points are closed. In the open position there should be a minimum gap of 0.018'' between the points.

## 5   Voltage regulator - checking the settings

1   To check the electrical settings it is necessary to have a good quality voltmeter of the moving coil type, with a range of 0 - 20 volts d.c. Connect the negative lead of the voltmeter to the D terminal of the regulator unit and the positive lead to the E terminal. Detach the negative lead from the battery (positive lead, early machines with negative earth). Start the engine and slowly increase the engine speed until the voltmeter needle 'kicks' then steadies. Note the reading and stop the engine.

2.1 Ammeter will normally give indication of dynamo output

4.1 Voltage regulator unit has two separate solenoids within

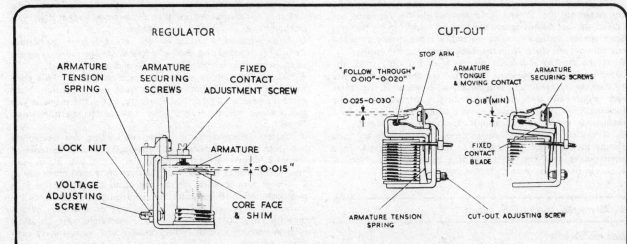

REGULATOR                                                    CUT-OUT

ARMATURE
TENSION
SPRING

ARMATURE
SECURING
SCREWS

FIXED
CONTACT
ADJUSTMENT SCREW

LOCK NUT

ARMATURE

VOLTAGE
ADJUSTING
SCREW

CORE FACE
& SHIM

=0·015"

"FOLLOW THROUGH"
0·010"–0·020"

STOP ARM

ARMATURE
TONGUE
& MOVING CONTACT

ARMATURE
SECURING SCREWS

0·025–0·030"

0·018"(MIN)

FIXED
CONTACT
BLADE

ARMATURE TENSION
SPRING

CUT-OUT ADJUSTING SCREW

Fig. 8.1 Regulator and cut-out adjustment and setting

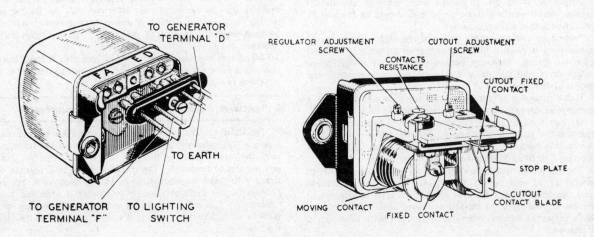

TO GENERATOR
TERMINAL "D"

TO EARTH

TO GENERATOR
TERMINAL "F"

TO LIGHTING
SWITCH

REGULATOR ADJUSTMENT
SCREW

CONTACTS
RESISTANCE

CUTOUT ADJUSTMENT
SCREW

CUTOUT FIXED
CONTACT

STOP PLATE

CUTOUT
CONTACT BLADE

MOVING CONTACT

FIXED CONTACT

Fig. 8.2 Control box connections and internal layout

2  The setting is correct if the voltage on open circuit is within the following limits:

| Air temperature | | Acceptable voltage range |
|---|---|---|
| 10°C | 50°F | 7.7 — 8.1 |
| 20°C | 68°F | 7.6 — 8.0 |
| 30°C | 86°F | 7.5 — 7.9 |
| 40°C | 104°F | 7.4 — 7.8 |

If the voltage is outside the acceptable range, the regulating screw must be readjusted. Turn clockwise to raise the setting, or anti-clockwise to decrease it, noting that only a fraction of a turn may be necessary to achieve a marked change in the readings. When the setting is correct, tighten the locknut.

3  Adjustment should be effected within 30 seconds, otherwise the shunt winding will heat up and give rise to false settings. If the regulator unit is removed during these checking operations, make sure it is held in a similar position to that adopted on the machine.

4  Do not run the engine at more than half engine speed, otherwise the dynamo will build up a high voltage because it is running on open circuit.

5  To check the electrical setting of the cut-out, connect the voltmeter to the D and E terminals of the regulator, as previously, but do not detach the negative lead from the battery. Start the engine and increase the engine speed slowly until the cut-out contacts close. Note the reading and stop the engine.

6  If the reading is outside the limits of 6.3 - 6.7 volts, the cut-out adjusting screw must be reset. Turn the screw clockwise to increase the reading or anti-clockwise to reduce it. Test after each adjustment and when the reading is correct, tighten the locknut. The cut-out adjusting screw is equally sensitive and false readings are again liable to occur if adjustment is prolonged.

7  If the cut-out points fail to close, there may be an open circuit in the regulator unit or in the dynamo itself. It is advisable to seek the assistance of a qualified auto-electrical expert at this stage.

## 6  Alternator - checking the output

### Late models only

1  the output and performance of the alternator fitted to the late type pre-unit construction twins can be checked only with specialised test equipment of the multi-meter type. It is unlikely that the average owner will have access to this type of equipment or instruction in its use. In consequence, if the performance is suspect, the alternator and charging circuit should be checked by a qualified auto-electrical expert.

2  Failure of the alternator does not necessarily mean that a replacement is needed. This can however sometimes be most economic through a service exchange scheme. It is possible to replace or rewind the stator coil assembly, for example, if the rotor is undamaged.

3  If the generator fails to charge, the ammeter will show only a discharge which will increase when the lights are switched on.

4  Do not omit to check the rectifier before pronouncing the alternator at fault. If this component has failed, it will give identical symptoms. There is no way of reconditioning or repairing a rectifier. If it malfunctions, it must be renewed.

5  The early alternators are of the six volt type. It is possible to convert to 12 volt operation, but only with the appropriate parts, obtainable from a Lucas agent. This type of conversion can be applied only to models equipped with an alternator. Don't forget the battery, horn and all the light bulbs have also to be changed.

## 7  Battery - charging procedure and maintenance

1  Whilst the machine is used on the road it is unlikely that the battery will require attention other than routine maintenance because the generator will keep it fully charged. However, if the machine is used for a succession of short journeys only, mainly during the hours of darkness when the lights are in full use, it is possible that the output from the generator may fail to keep pace with the heavy electrical demand, especially if the machine is parked with the lights switched on. Under these circumstances, it will be necessary to remove the battery from time to time to have it charged independently.

2  The battery is located below the dual seat, in a carrier. It is secured by a strap which, when released, will permit the battery to be withdrawn after disconnection of the leads. The battery positive is always earthed. To remove the battery release the carrier clamp and disconnect the electrical leads. Later models may have a different form of battery holder, although the battery will be found in approximately the same location below the dual seat.

3  The normal charge rate is 1 amp. A more rapid charge can be given in an emergency, but this should be avoided if possible because it will shorten the life of the battery.

4  When the battery is removed from the machine, remove the cover and clean the battery top. If the terminals are corroded, scrape them clean and cover them with vaseline (not grease) to protect them from further attack. If a vent tube is fitted, make sure it is not obstructed and that it is arranged so that it will not discharge over any parts of the machine.

5  If the machine is laid up for any period of time, the battery should be removed and given a 'refresher' charge every six weeks or so, in order to maintain it in good condition.

6  Several different types of Lucas lead-acid battery have been fitted to the twin cylinder models since their inception. Most batteries have a 13 amp. hour capacity, but if the machine was specified for sidecar use, a 22 amp. hour battery was sometimes fitted.

7  Battery maintenance is limited to keeping the electrolyte level just above the plates and separators. Modern batteries of the MLZ9E type have a transparent plastics case with a blue level line, which makes the check of the electrolyte level much easier.

8  Unless acid is spilt, which may occur if the machine falls over, use only distilled water for topping up purposes, until the correct level is restored. If acid is split on any part of the machine, it should be neutralised immediately with an alkali such as washing soda or baking powder, and washed away with plenty of water. This will prevent corrosion from taking place. Top up in this instance with sulphuric acid of the correct specific gravity (1.260 - 1.280).

9  It is seldom practicable to repair a cracked battery case because the acid that is already seeping through the crack will prevent the formation of an effective seal, no matter what sealing compound is used. It is always best to replace a cracked battery, especially in view of the risk of corrosion from the acid leakage.

10  Make sure the battery is clamped securely. A loose battery will vibrate and its working life will be greaty shortened, due to the paste being shaken out of the plates.

## 8  Headlamp - replacing bulbs and adjusting beam height

1  Pre-1948 models are fitted with a headlamp that has a separate rim, glass and reflector unit. To remove the front, press back the retaining clip at the base of the headlamp shell and lift out the rim and reflector. The main headlamp bulb and the pilot lamp bulb have bayonet fittings that engage with the bulb holders of the reflector unit. The former is attached to the reflector by two retaining springs and contains provision for the bulb to be focussed by moving it either backward or forward in the holder.

2  To replace the front, locate the top of the rim first, then press the bottom of the combined front and reflector unit so that the retaining clip can be re-engaged.

3  1948 models employ a somewhat similar method of fixing, except that the retaining clip at the bottom of the shell takes the form of a spring-loaded catch which must be pulled to release the rim. The main bulb holder slides within an extension of the

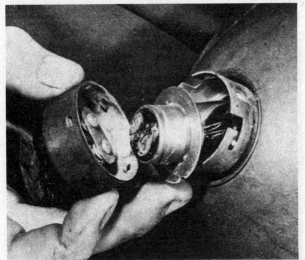
8.1 Pre-focus headlamp bulb fits into rear of reflector unit

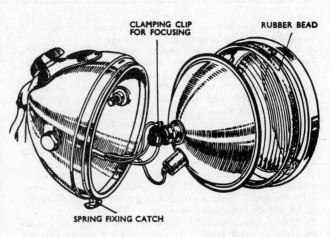
Fig. 8.3 Headlamp construction (pre 1948)

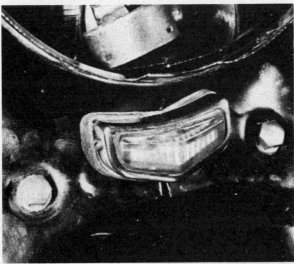
8.4 Some machines have lower, underslung pilot bulb

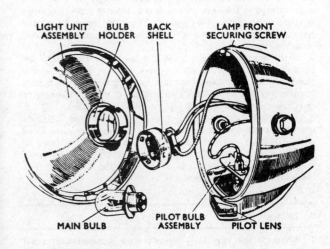
Fig. 8.4 Headlamp construction (pre-focus type)

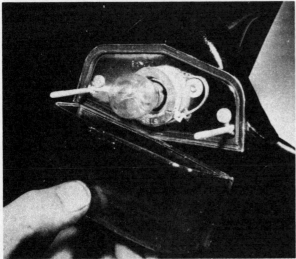
9.2 Remove plastic lens for access to tail/stop lamp bulb

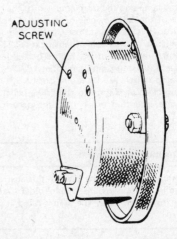
Fig. 8.5 Typical Lucas electric horn, showing adjusting screw

reflector unit to provide facilities for focussing. It is locked in position by means of a metal clamp and screw.

4 Later models have an integral headlamp, in which the reflector and the headlamp glass are sealed together. This type of lamp rim is secured by a screw in the top of the headlamp shell, which must be slackened before the rim complete with lamp unit can be drawn off. The main headlamp bulb is of the pre-focus type, which has a special type of bayonet fitting connector. A rim that is integral with the bulb ensures the bulb can be replaced in only the correct position; earlier bulbs have the bayonet cap marked 'TOP' so that the main and dip contacts cannot be transposed inadvertently.

5 Beam height is adjusted by slackening the two bolts that retain the headlamp shell to the forks and tilting the headlamp either upwards or downwards. Adjustments should always be made with the rider seated normally.

6 U.K. lighting regulations stipulate that the lighting system must be arranged so that the light does not dazzle a person standing in the same horizontal plane as the vehicle, at a distance greater than 25 yards from the lamp, whose eye level is not less than 3 feet six inches above that plane. It is easy to approximate this setting by placing the machine 25 yards away from a wall, on a level road and setting the beam height so that it is concentrated at the same height as the distance from the centre of the headlamp to the ground. The rider must be seated normally during this operation, and the pillion passenger, if one is carried regularly.

7 If the headlamp bulb is broken, it can be removed from the headlamp rim by detaching the wire retaining clips, after the front has been removed from the headlamp. In the case of a headlamp of the unit glass/reflector type, it will be necessary to purchase the complete beam unit, and not the glass alone.

8 The main headlamp bulb has a 30/30W rating (24/30W pre-focus bulb) except in the case of the earlier models fitted with the lower output Lucas E3H dynamo. A bulb rated at 24/24W is then specified. For all types of headlamp, a pilot lamp bulb of 6W rating is specified to meet with the requirements of the existing light regulations.

## 9 Tail and stop lamp - replacing bulbs

1 Although the very early models were supplied with a tail lamp only, it is doubtful whether any of the original fittings are still in use because the size no longer meets the minimum requirements of the lighting regulations. Most of the current tail lamp units contain provision for a stop lamp bulb also, which is operated when the rear brake pedal is depressed.

2 Removal of the plastic lens cover will reveal the bulb holders for both the tail lamp and the stop lamp, which may be separate, as in the case of the early models or combined, to conform to current practice. It is now customary to fit a single bulb with offset pins, which is of the double filament type. The offset pins prevent accidental inversion of the bulb. The tail lamp filament is rated at 6W and the stop lamp filament at 18W.

3 The stop lamp switch will be found on the left-hand side of the machine, in close proximity to the brake pedal. The switch does not require attention other than the occasional drop of light oil.

## 10 Speedometer bulb - replacement

1 The bulb that illuminates the dial of this instrument has a bayonet fitting in a rubber-mounted bulbholder that screws into the base of the instrument case. The bulb is rated at 2.2W.

## 11 Horn - adjustment

1 A horn of the electro-magnetic type is fitted to every

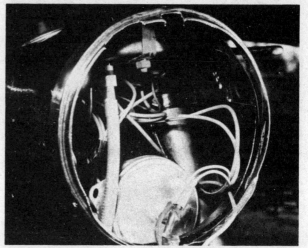
11.1 Horn is contained with nacelle unit

11.1a Adjustments are made at the rear of the horn body

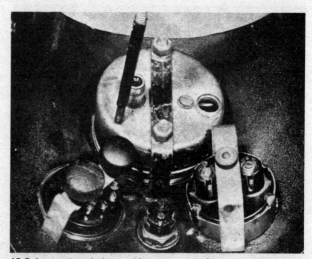
13.2 Access to switches and instruments is from underside of nacelle top

machine. It is operated from a push button, mounted on the handlebars.

2  A small, serrated screw, located in the back of the horn body, provides means of adjusting the note. It must not be turned more than two or three notches, before re-testing. If the ammeter shows a reading when the horn is operated, it should not exceed 4 amps. It is possible to obtain a good note, at the expense of an excessive current requirement.

## 12 Wiring - layout and examination

1  The cables of the wiring harness are colour-coded and will correspond with the accompanying wiring diagrams.

2  Visual inspection will show whether any breaks or frayed outer coverings are giving rise to short circuits which will cause the main fuse to blow. Another source of trouble is the snap connectors and spade terminals, which may make a poor connection if they are not pushed home fully.

3  Intermittent short circuits can sometimes be traced to a chafed wire passing through, or close to, a metal component, such as a frame member. Avoid tight bends in the cables or situations where the cables can be trapped or stretched, especially in the vicinity of the handlebars or steering head.

## 13 Headlamp switch

1  It is unusual for the headlamp switch to give trouble unless the machine has been laid up for a considerable period and the switch contacts have become dirty. Contact between the terminal posts is made by a spring-loaded roller attached to the body of the switch knob. If the terminal posts become corroded or oxidised, poor or intermittent electrical contact will result.

2  It is possible to dismantle the switch and clean the terminal posts and roller by hand; the rotor containing the roller can be pulled away when the centre screw of the switch knob is removed and the knob detached. This is a somewhat delicate operation, which should be performed only when the switch complete is removed from the headlamp shell. The switch body is held to the underside of the shell by means of a wire spring that engages with a groove around the body moulding.

3  A better alternative that does not necessitate dismantling the switch is the use of one of the proprietary switch contact cleansers that are available in aerosol form.

4  On no account oil the switch or oil will spread across the internal contacts, to form an effective insulator.

## 14 Fault diagnosis - electrical system

| Symptom | Cause | Remedy |
| --- | --- | --- |
| Complete electrical failure | Short circuit | Check wiring and electrical components for insulation breakdown. |
|  | Isolated battery | Check battery connections, also whether connections show signs of corrosion. |
| Dim lights, horn inoperative | Discharged battery | Recharge battery with battery charger and check whether alternator is giving correct output. |
| Constantly 'blowing' bulbs | Vibration, poor earth connection | Check whether bulb holders are secured correctly. Check earth return or connections to frame. |

**Wiring diagrams commence overleaf**

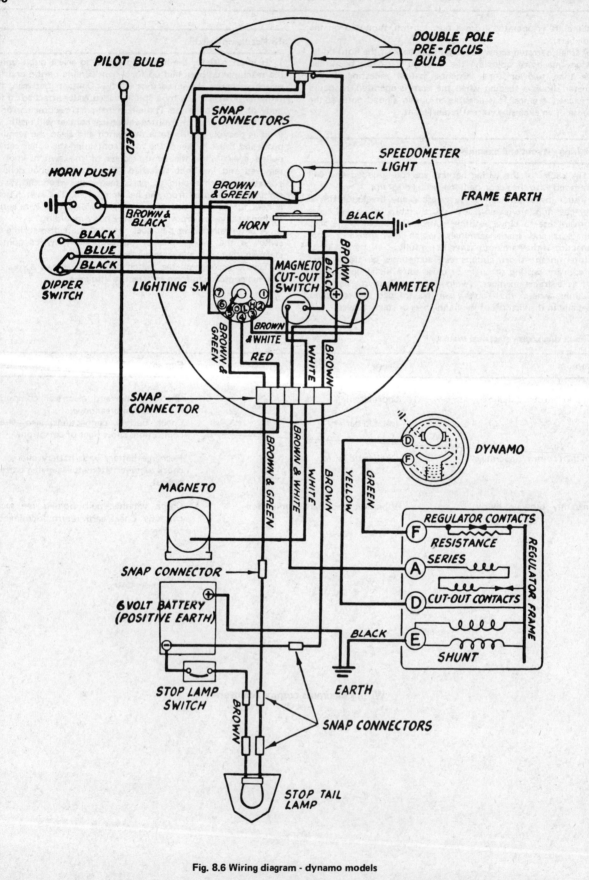

**Fig. 8.6 Wiring diagram - dynamo models**

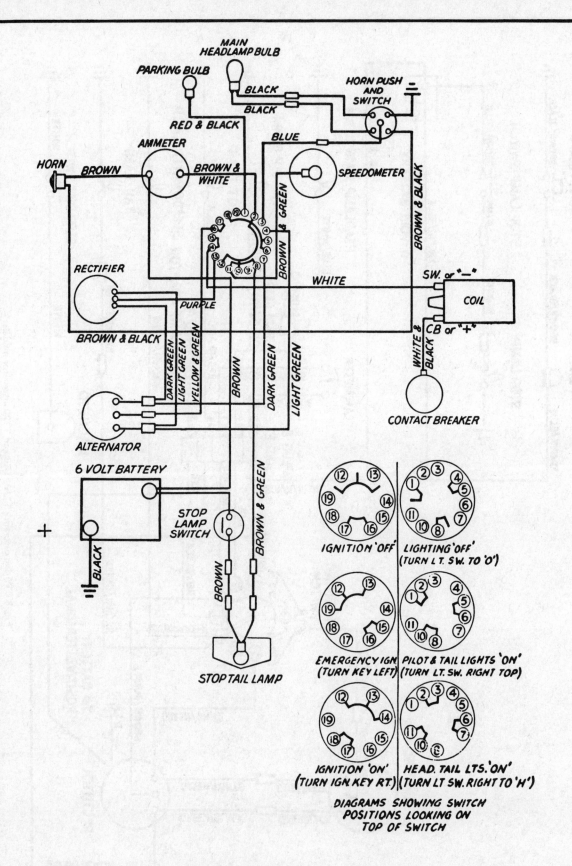

Fig. 8.7 Wiring diagram - 5T and 6T alternator models

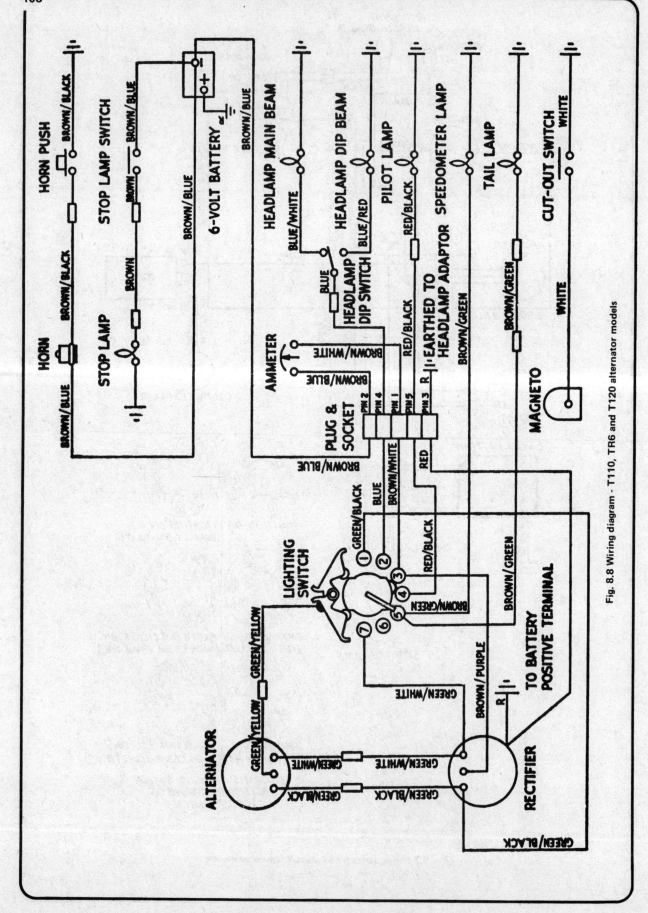

Fig. 8.8 Wiring diagram - T110, TR6 and T120 alternator models

# Metric conversion tables

| Inches | Decimals | Millimetres | Millimetres to Inches | | Inches to Millimetres | |
|--------|----------|-------------|------|--------|--------|------|
| | | | mm | Inches | Inches | mm |
| 1/64 | 0.015625 | 0.3969 | 0.01 | 0.00039 | 0.001 | 0.0254 |
| 1/32 | 0.03125 | 0.7937 | 0.02 | 0.00079 | 0.002 | 0.0508 |
| 3/64 | 0.046875 | 1.1906 | 0.03 | 0.00118 | 0.003 | 0.0762 |
| 1/16 | 0.0625 | 1.5875 | 0.04 | 0.00157 | 0.004 | 0.1016 |
| 5/64 | 0.078125 | 1.9844 | 0.05 | 0.00197 | 0.005 | 0.1270 |
| 3/32 | 0.09375 | 2.3812 | 0.06 | 0.00236 | 0.006 | 0.1524 |
| 7/64 | 0.109375 | 2.7781 | 0.07 | 0.00276 | 0.007 | 0.1778 |
| 1/8 | 0.125 | 3.1750 | 0.08 | 0.00315 | 0.008 | 0.2032 |
| 9/64 | 0.140625 | 3.5719 | 0.09 | 0.00354 | 0.009 | 0.2286 |
| 5/32 | 0.15625 | 3.9687 | 0.1 | 0.00394 | 0.01 | 0.254 |
| 11/64 | 0.171875 | 4.3656 | 0.2 | 0.00787 | 0.02 | 0.508 |
| 3/16 | 0.1875 | 4.7625 | 0.3 | 0.01181 | 0.03 | 0.762 |
| 13/64 | 0.203125 | 5.1594 | 0.4 | 0.01575 | 0.04 | 1.016 |
| 7/32 | 0.21875 | 5.5562 | 0.5 | 0.01969 | 0.05 | 1.270 |
| 15/64 | 0.234375 | 5.9531 | 0.6 | 0.02362 | 0.06 | 1.524 |
| 1/4 | 0.25 | 6.3500 | 0.7 | 0.02756 | 0.07 | 1.778 |
| 17/64 | 0.265625 | 6.7469 | 0.8 | 0.03150 | 0.08 | 2.032 |
| 9/32 | 0.28125 | 7.1437 | 0.9 | 0.03543 | 0.09 | 2.286 |
| 19/64 | 0.296875 | 7.5406 | 1 | 0.03947 | 0.1 | 2.54 |
| 5/16 | 0.3125 | 7.9375 | 2 | 0.07874 | 0.2 | 5.08 |
| 21/64 | 0.328125 | 8.3344 | 3 | 0.11811 | 0.3 | 7.62 |
| 11/32 | 0.34375 | 8.7312 | 4 | 0.15748 | 0.4 | 10.16 |
| 23/64 | 0.359375 | 9.1281 | 5 | 0.19685 | 0.5 | 12.70 |
| 3/8 | 0.375 | 9.5250 | 6 | 0.23622 | 0.6 | 15.24 |
| 25/64 | 0.390625 | 9.9219 | 7 | 0.27559 | 0.7 | 17.78 |
| 13/32 | 0.40625 | 10.3187 | 8 | 0.31496 | 0.8 | 20.32 |
| 27/64 | 0.421875 | 10.7156 | 9 | 0.35433 | 0.9 | 22.86 |
| 7/16 | 0.4375 | 11.1125 | 10 | 0.39370 | 1 | 25.4 |
| 29/64 | 0.453125 | 11.5094 | 11 | 0.43307 | 2 | 50.8 |
| 15/32 | 0.46875 | 11.9062 | 12 | 0.47244 | 3 | 76.2 |
| 31/64 | 0.484375 | 12.3031 | 13 | 0.51181 | 4 | 101.6 |
| 1/2 | 0.5 | 12.7000 | 14 | 0.55118 | 5 | 127.0 |
| 33/64 | 0.515625 | 13.0969 | 15 | 0.59055 | 6 | 152.4 |
| 17/32 | 0.53125 | 13.4937 | 16 | 0.62992 | 7 | 177.8 |
| 35/64 | 0.546875 | 13.8906 | 17 | 0.66929 | 8 | 203.2 |
| 9/16 | 0.5625 | 14.2875 | 18 | 0.70866 | 9 | 228.6 |
| 37/64 | 0.578125 | 14.6844 | 19 | 0.74803 | 10 | 254.0 |
| 19/32 | 0.59375 | 15.0812 | 20 | 0.78740 | 11 | 279.4 |
| 39/64 | 0.609375 | 15.4781 | 21 | 0.82677 | 12 | 304.8 |
| 5/8 | 0.625 | 15.8750 | 22 | 0.86614 | 13 | 330.2 |
| 41/64 | 0.640625 | 16.2719 | 23 | 0.90551 | 14 | 355.6 |
| 21/32 | 0.65625 | 16.6687 | 24 | 0.94488 | 15 | 381.0 |
| 43/64 | 0.671875 | 17.0656 | 25 | 0.98425 | 16 | 406.4 |
| 11/16 | 0.6875 | 17.4625 | 26 | 1.02362 | 17 | 431.8 |
| 45/64 | 0.703125 | 17.8594 | 27 | 1.06299 | 18 | 457.2 |
| 23/32 | 0.71875 | 18.2562 | 28 | 1.10236 | 19 | 482.6 |
| 47/64 | 0.734375 | 18.6531 | 29 | 1.14173 | 20 | 508.0 |
| 3/4 | 0.75 | 19.0500 | 30 | 1.18110 | 21 | 533.4 |
| 49/64 | 0.765625 | 19.4469 | 31 | 1.22047 | 22 | 558.8 |
| 25/32 | 0.78125 | 19.8437 | 32 | 1.25984 | 23 | 584.2 |
| 51/64 | 0.796875 | 20.2406 | 33 | 1.29921 | 24 | 609.6 |
| 13/16 | 0.8125 | 20.6375 | 34 | 1.33858 | 25 | 635.0 |
| 53/64 | 0.828125 | 21.0344 | 35 | 1.37795 | 26 | 660.4 |
| 27/32 | 0.84375 | 21.4312 | 36 | 1.41732 | 27 | 685.8 |
| 55/64 | 0.859375 | 21.8281 | 37 | 1.4567 | 28 | 711.2 |
| 7/8 | 0.875 | 22.2250 | 38 | 1.4961 | 29 | 736.6 |
| 57/64 | 0.890625 | 22.6219 | 39 | 1.5354 | 30 | 762.0 |
| 29/32 | 0.90625 | 23.0187 | 40 | 1.5748 | 31 | 787.4 |
| 59/64 | 0.921875 | 23.4156 | 41 | 1.6142 | 32 | 812.8 |
| 15/16 | 0.9375 | 23.8125 | 42 | 1.6535 | 33 | 838.2 |
| 61/64 | 0.953125 | 24.2094 | 43 | 1.6929 | 34 | 863.6 |
| 31/32 | 0.96875 | 24.6062 | 44 | 1.7323 | 35 | 889.0 |
| 63/64 | 0.984375 | 25.0031 | 45 | 1.7717 | 36 | 914.4 |

# English/American terminology

Because this book has been written in England, British English component names, phrases and spellings have been used throughout. American English usage is quite often different and whereas normally no confusion should occur, a list of equivalent terminology is given below.

| English | American | English | American |
|---|---|---|---|
| Air filter | Air cleaner | Number plate | License plate |
| Alignment (headlamp) | Aim | Output or layshaft | Countershaft |
| Allen screw/key | Socket screw/wrench | Panniers | Side cases |
| Anticlockwise | Counterclockwise | Paraffin | Kerosene |
| Bottom/top gear | Low/high gear | Petrol | Gasoline |
| Bottom/top yoke | Bottom/top triple clamp | Petrol/fuel tank | Gas tank |
| Bush | Bushing | Pinking | Pinging |
| Carburettor | Carburetor | Rear suspension unit | Rear shock absorber |
| Catch | Latch | Rocker cover | Valve cover |
| Circlip | Snap ring | Selector | Shifter |
| Clutch drum | Clutch housing | Self-locking pliers | Vise-grips |
| Dip switch | Dimmer switch | Side or parking lamp | Parking or auxiliary light |
| Disulphide | Disulfide | Side or prop stand | Kick stand |
| Dynamo | DC generator | Silencer | Muffler |
| Earth | Ground | Spanner | Wrench |
| End float | End play | Split pin | Cotter pin |
| Engineer's blue | Machinist's dye | Stanchion | Tube |
| Exhaust pipe | Header | Sulphuric | Sulfuric |
| Fault diagnosis | Trouble shooting | Sump | Oil pan |
| Float chamber | Float bowl | Swinging arm | Swingarm |
| Footrest | Footpeg | Tab washer | Lock washer |
| Fuel/petrol tap | Petcock | Top box | Trunk |
| Gaiter | Boot | Torch | Flashlight |
| Gearbox | Transmission | Two/four stroke | Two/four cycle |
| Gearchange | Shift | Tyre | Tire |
| Gudgeon pin | Wrist/piston pin | Valve collar | Valve retainer |
| Indicator | Turn signal | Valve collets | Valve cotters |
| Inlet | Intake | Vice | Vise |
| Input shaft or mainshaft | Mainshaft | Wheel spindle | Axle |
| Kickstart | Kickstarter | White spirit | Stoddard solvent |
| Lower leg | Slider | Windscreen | Windshield |
| Mudguard | Fender | | |

# Index